A Book Of

LASERS

T.Y.B.Sc. Physics : PH – 346 (K) : Semester-IV
As Per New Revised Syllabus With Effect from June 2015

Dr. A. B. BHISE
(M.Sc., M. Phil., Ph.D)
Ex-Head
Department of Physics,
C.T. Bora College, Shirur,
Tal. : Shirur, Dist. : **PUNE**.

R. B. BHISE
(M.Sc., M.Phil., D.C.M.)
Head
Department of Physics,
B.J.A.C.S. College, Ale
Tal. : Junnar, Dist. : **PUNE**.

Dr. S. M. RATHOD
(M.Sc., M. Phil., Ph.D)
Associate Professor,
Department of Physics,
Abasaheb Garware College,
PUNE.

S. F. DHAKANE
(M.Sc., M.Phil.)
Head
Department of Physics,
A. W. College, Otur
Tal. : Junnar, Dist. : **PUNE**

N1869

T.Y.B.Sc. : INTRODUCTION TO LASERS (S-IV)　　　　**ISBN 978-93-5164-926-7**

Third Edition : **January 2020**

© : **Authors**

Published By :
NIRALI PRAKASHAN
Abhyudaya Pragati, 1312, Shivaji Nagar,
Off J.M. Road, PUNE – 411005
Tel - (020) 25512336/37/39, Fax - (020) 25511379
Email : niralipune@pragationline.com

➢ DISTRIBUTION CENTRES

PUNE

Nirali Prakashan : 119, Budhwar Peth, Jogeshwari Mandir Lane, Pune 411002, Maharashtra
(For orders within Pune)　Tel : (020) 2445 2044, Mobile : 9657703145
　Email : niralilocal@pragationline.com

Nirali Prakashan : S. No. 28/27, Dhayari, Near Asian College Pune 411041
(For orders outside Pune)　Tel : (020) 24690204 Fax : (020) 24690316; Mobile : 9657703143
　Email : bookorder@pragationline.com

MUMBAI

Nirali Prakashan : 385, S.V.P. Road, Rasdhara Co-op. Hsg. Society Ltd.,
　Girgaum, Mumbai 400004, Maharashtra; Mobile : 9320129587
　Tel : (022) 2385 6339 / 2386 9976, Fax : (022) 2386 9976
　Email : niralimumbai@pragationline.com

➢ DISTRIBUTION BRANCHES

JALGAON

Nirali Prakashan : 34, V. V. Golani Market, Navi Peth, Jalgaon 425001, Maharashtra,
　Tel : (0257) 222 0395, Mob : 94234 91860;
　Email : niralijalgaon@pragationline.com

KOLHAPUR

Nirali Prakashan : New Mahadvar Road, Kedar Plaza, 1st Floor Opp. IDBI Bank, Kolhapur 416 012
　Maharashtra. Mob : 9850046155; Email : niralikolhapur@pragationline.com

NAGPUR

Nirali Prakashan : Above Maratha Mandir, Shop No. 3, First Floor,
　Rani Jhanshi Square, Sitabuldi, Nagpur 440012, Maharashtra
　Tel : (0712) 254 7129; Email : niralinagpur@pragationline.com

DELHI

Nirali Prakashan : 4593/15, Basement, Agarwal Lane, Ansari Road, Daryaganj
　Near Times of India Building, New Delhi 110002 Mob : 08505972553
　Email : niralidelhi@pragationline.com

BENGALURU

Nirali Prakashan : Maitri Ground Floor, Jaya Apartments, No. 99, 6th Cross, 6th Main,
　Malleswaram, Bengaluru 560003, Karnataka; Mob : 9449043034
　Email: niralibangalore@pragationline.com

Other Branches : Hyderabad, Chennai

niralipune@pragationline.com | www.pragationline.com
Also find us on www.facebook.com/niralibooks

Dedicated

to

My Father
Baburao

Preface ...

We are happy to place this book **"LASERS"** in the hand of T.Y.B.Sc students and teachers of Pune University. This book envisages the revised syllabus which is implemented from the academic year 2015-16.

This book gives a brief description of Lasers (Physics) which is as per the latest curriculum. The theory is presented in brief with the principle wherever necessary and its explanation.

At the end of each chapter necessary questions are included. We think that the students and teachers will find this book most useful while studying in B.Sc.

We are thankful to Mr. Ashok Bodake and his group members for their efforts and encouragement. We are also thankful to Dnyaneshear Gramonnati Mandal, B. J. College, Ale (PUNE), C. T. Bora College, Shirur, A. W. College, Otur and Abasaheb Garware College, Pune for their co-operation and encouragement. We must gratefully acknowledge the patience and understanding shown by our family members during this endeavor.

Our sincere thanks are to Hon. Dinesh Furia and whole staff of Nirali Prakashan for their co-operation, nice printing, constant encouragement and efforts in bringing out this book.

Any suggestion for improvement of the book shall be most welcome.

NOVEMBER 2015   **AUTHORS**

PUNE

Syllabus ...

1. Introduction to Lasers : (08 L)

    1.1   Ordinary light and Lasers

    1.2   Brief history of Laser

    1.3   Interaction of radiation with matter

    1.4   Energy levels

    1.5   Population density

    1.6   Boltzmann distribution

    1.7   Transition Lifetimes

    1.8   Allowed and Forbidden Transitions

    1.9   Stimulated Absorption, Spontaneous Emission and Stimulated Emission

    1.10 Einstein's Coefficients and Einstein's relations

2. Laser Action : (06 L)

    2.1   Condition for large stimulated emission,

    2.2   Population inversion

    2.3   Condition for light amplification,

    2.4   Gain coefficient, Active medium, Metastable states

    2.5   Pumping schemes : Three-level and four-level

3. Laser Oscillator : (07 L)

    3.1   Optical feedback

    3.2   Round trip gain

    3.3   Threshold gain

    3.4   Critical population inversions

    3.5   Optical resonator

    3.6   Condition for steady state oscillations

    3.7   Cavity resonance frequencies

4. Laser Output : (03 L)

    4.1   Lineshape broadening

    4.2   Lifetime broadening

    4.3   Collision broadening

    4.4   Doppler broadening

5. Characteristics of Lasers : (04 L)

    5.1   Directionality

    5.2   Monochromaticity

    5.3   Coherence

    5.4   Brightness

6. Types of Lasers : (12 L)

    6.1   Solid State Lasers : Ruby Laser

    6.2   Gas Lasers : He-Ne Laser, CO_2 Laser

    6.3   Liquid Lasers : Tunable dye laser

    6.4   Semiconductor Diode Laser

7. Applications of Lasers : (08 L)

    7.1   Industrial : Welding, cutting, drilling

    7.2   Nuclear Science : Laser isotope separation, laser fusion,

    7.3   Defense : Range finder

    7.4   Medical : Eye surgery

    7.5   Optical : Holography, super market scanners, compact discs

❑❑❑

Contents ...

❑❑❑

Chapter **1** ...

Introduction to Lasers

Objectives ...

- To study difference between ordinary light and lasers.

- To understand brief history of lasers.

- To discuss interaction of radiation with matter, Energy levels, Population density, and Boltzmann distribution.

- To discuss the fundamental terms such as absorption, spontaneous and stimulated emission.

- To study and derive Einstein's expression for stimulated emission.

1.1 Ordinary Light and Lasers　　　　　　　　　(April 2017, 2018)

Photons that make up ordinary light are characterized by their energy (wavelength), direction of travel, and polarization (direction of oscillating electric field). The difference between ordinary light and laser light is :

1. Ordinary light is polychromatic, means it consists of all the constituent colours of light. The ordinary light contains wavelengths roughly in the range of 400-700 nanometers. Whereas, light from a laser is monochromatic, means it consists of radiations of only one wavelength.

2. Ordinary light is mostly unpolarized; whereas, light from a laser is polarized, means all radiations are oscillating in the same direction. Hence, laser light is highly directional.

3. The basic difference is that the ordinary sources are incoherent, that means the discrete frequencies merge up to give an intermediate between the maximum and minimum frequencies; whereas, the laser light is highly coherent containing the single frequency with maximum amplitude and can be focused to a very small spot. Thus laser light can travel far away.

4. Intensity of ordinary light decreases with distance, as it travels in the form of short pulses of small length and short duration; whereas, laser light travels in only one direction, which is parallel to the optic axis, hence, laser light is highly intense.

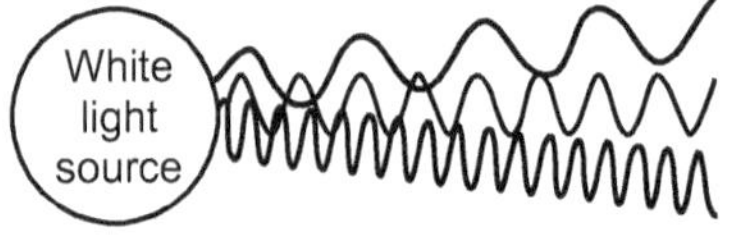

(a) Path of white light   **(b) Path of laser light**

Fig. 1.1

Laser light is a highly directional, monochromatic, polarized and high intensity beam with a narrow frequency range than that available from common types of sources. Therefore, it is used in industries, medical, military, photography, etc.

1.2 Brief History of Lasers   (April 2017)

The development of Lasers has been proved to be a turning point in the history of science and engineering. It has produced a completely new type of systems with potentials for applications in a wide variety of fields. Lasers are a highly directional, monochromatic, highly coherent and high intensity beams. Laser is based on amplification of light. Laser is an acronym for Light Amplification by Stimulated Emission of Radiation. The brief history of laser is as follows.

In 1900, Max Plank published work that provided the understanding that light is a form of electromagnetic radiation. In 1917, the principle of the laser was first discovered, when physicist Albert Einstein described the theory of stimulated emission. He postulated that, when the population inversion exists between upper and lower levels among atomic systems, it is possible to realize amplified stimulated emission and the stimulated emission has the same frequency and phase as the incident radiation. In 1951, Charles H. Townes and his co-workers independently investigated and suggested the principle of MASER device (Microwave Amplification of Stimulated Emission of Radiation) based on stimulated emission, at Columbia University and Lebedev Institute of Physics, Moscow. MASER is a device that amplifies microwaves for its immediate application in microwave communication systems. In 1960, C. H. Townes had patented for laser and he was awarded by Nobel Prize in 1964. Despite the pioneering work of Townes and Prokhorov it was left to Theodore Maiman, on 16th May 1960, he was invented first working LASER based on Ruby crystal as a lasing medium that was stimulated using high energy flashes of intense light in Hughes Research Laboratories. *In December 1960, Ali Javan and his co-worker invented first Helium-Neon gas LASER at Bell Laboratory.*   **(April 2016, Nov. 2016)**

During sixties, lot of work had been carried out on the basic development of almost all the major lasers including high power gas dynamic and chemical lasers. Almost all the practical applications of these lasers in defense as well as in industry were also identified during this period. The motivation of using the high power lasers in strategic scenario was a great driving force for the rapid development of these high power lasers. In early seventies, megawatt class carbon dioxide gas dynamic laser was successfully developed and tested against typical military targets. The development of chemical lasers, free electron and X-ray lasers took slightly longer time because of involvement of multidisciplinary approach.

The major steps of advances in Laser research are given below :

1962　:　Hall R. - First Semiconductor laser (Gallium-Arsenide laser) at General Electric Labs.

1962　:　Hellwarth, R.W. - Giant pulse generation / Q-Switching.

1962　:　Johnson L.F. - Continuous wave solid-state laser.

1964　:　Geusic J.E. - Development of first working Nd :YAG LASER at Bell Labs.

1964　:　Patel C.K.N. - Development of CO_2 LASER at Bell Labs.

1964　:　Bridges W. - Development of Argon Ion LASER at Hughes Labs.

1965　:　Kasper J. V. V. - First chemical LASER at University of California, Berkley.

1966　:　Fowles, G. - First metal vapour LASER - Zn/Cd - at University of Utah.

1966　:　Sorokin P. - Demonstration of first Dye Laser action at IBM Labs.

1977　:　McDermott W.E. - Chemical Oxygen Iodine Laser (COIL).

2001　:　Lawrence Livermore National Laboratory - Solid State Heat Capacity Laser (SSHCL).

1.3 Interaction of Radiation with Matter

Interaction of electromagnetic radiation with matter produces absorption and spontaneous emissions. Absorption and spontaneous emissions are natural processes. For the generation of laser, stimulated emission is essential. Stimulated emission has to be induced or stimulated and is generated under special conditions as stated by Einstein in 1917. Absorption of electromagnetic radiation is the way in which the energy of a photon is taken up by matter. The reduction in intensity of a light wave propagating through a medium by absorption of a part of its photons is often called attenuation. The absorption of waves does not depend on their intensity.

The frequency (υ) and the wavelength (λ) of light are related to velocity of light (c) by the relation　　　　　　　$c = \upsilon\lambda$　　　　　　　... (1.1)

When a beam of light passes through absorbing medium, then the attenuation of light in medium describes the decrease in intensity of light (dI) (absorption) which is proportional to intensity of light and thickness of the medium.

$$dI = -\alpha\, I\, dx \qquad\qquad ... (1.2)$$

where, α is a coefficient of absorption and negative sign indicates that intensity decreases with respect to thickness of the medium.

$$\frac{dI}{I} = - \alpha \, dx$$

Integrating on both sides,

$$\int_{I_0}^{I} \frac{dI}{I} = - \alpha \int_{0}^{x} dx$$

$$I = I_0 \, e^{-\alpha x} \qquad \qquad \dots (1.3)$$

Above equation indicates that light intensity decreases exponentially with distance in the medium.

1.4 Energy Levels (April 2017)

The electron can gain the energy it needs by absorbing light. If the electron jumps from the second energy level down to the first energy level, it must give off some energy by emitting light. The atom absorbs or emits light in discrete packets called photons, and each photon has a definite energy. The energy that a photon carries depends on its wavelength. Since the photons absorbed or emitted by electrons jumping between the n = 1 and n = 2 energy levels must have exactly 10.2 eV of energy, the light absorbed or emitted must have a definite wavelength. This wavelength can be found from the equation

$$E = \frac{hc}{\lambda} \qquad \qquad \dots (1.4)$$

where E is the energy of the photon (in eV), h is Planck's constant (4.14×10^{-15} eV $= 6.64 \times 10^{-34}$ Js) and c is the speed of light (3×10^{8} m/s).

The electrons of individual atoms can be excited from a lower energy state to a higher energy state.

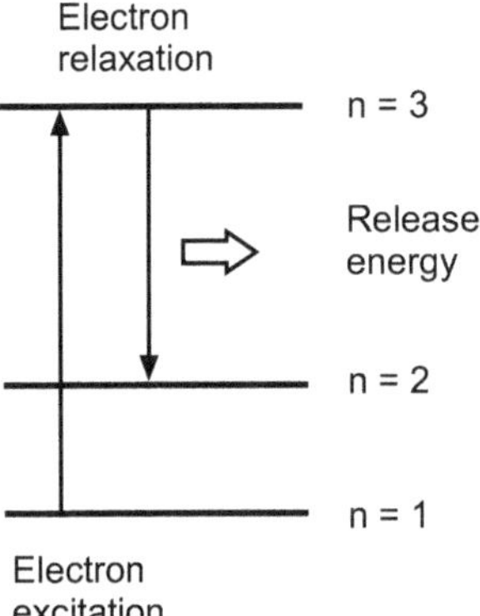

Fig. 1.2

This happens, for example, when a high voltage is applied through a collection of atoms in the gas phase. However, electrons tend to avoid excited states, so after a short time there is transition of electrons to a lower energy state (relaxation). Because energy was absorbed to excite the electrons, so energy has to be emitted after the transition of electrons to their initial state. This energy is released in the form of light (electromagnetic waves). Each energy level is labeled with the quantum number n (n = 1, 2, 3...) and the energy difference between the two energy levels involved in the transition. Niels Bohr proposed that electrons inside an atom occupy only certain allowed orbitals with a specific energy. The energy of an electron in an atom is not continuous, but 'quantized.' The energies corresponding to each of the allowed orbitals are called energy levels.

1.5 Population Density   (Oct. 2017, April 2017, Nov. 2016)

The production of a population inversion is a necessary step in the working of a standard laser. A situation not at equilibrium, means number of atoms N_2 in excited state is always more than number of atoms N_1 in lower state, can be created by adding energy via a process known as "pumping" in order to raise enough atoms to the upper level. Number of atoms per unit volume present in an energy level is called its population density. A population inversion occurs when a system exists in a state with more members in an excited state than in lower energy states.

1.6 Boltzmann Distribution

At thermodynamic equilibrium, the distribution of the atoms between the levels is given by Boltzmann's law. Assume there is a group of N atoms, each of which is capable of being in one of the two energy states, either

1. The ground state, with energy E_1; or

2. The excited state, with energy E_2, with $E_2 > E_1$.

The number of atoms which are in the ground state is given by N_1 and the number of atoms in the excited state is given by N_2. Then total number of atoms (N) is given by

$$N = N_1 + N_2$$

The energy difference between the two states is given by

$$\Delta E_{12} = E_2 - E_1 \qquad \qquad \ldots (1.5)$$

The characteristic frequency υ_{12} of light, which will interact with the atoms, is given by the relation

$$E_2 - E_1 = \Delta E = h\upsilon_{12} \qquad \qquad \ldots (1.6)$$

where, h is Planck's constant.

If the group of atoms is in thermal equilibrium, it can be shown from Maxwell-Boltzmann distribution that the ratio of the number of atoms in each state is given by the Boltzmann factor :   **(April 2016)**

$$N_2 = N_1 e^{-((E_2 - E_1)/KT)} \qquad \qquad \text{... (1.7)}$$

where T is the thermodynamic temperature of the group of atoms, and K is Boltzmann's constant.

We may calculate the ratio of the populations of the two states at room temperature (T $\approx$ 300 K) for an energy difference ΔE that corresponds to light of a frequency corresponding to visible light ($\upsilon \approx 5 \times 10^{14}$ Hz).

1.7 Transition Lifetimes

A laser transition is a transition of some laser-active ion between two electronic levels. It is also called amplifier transition, *for* example, where stimulated emission can take place and this leads to optical amplification. This amplification can be used in an optical amplifier or a laser. A long upper-state lifetime in a laser gain medium means that a significant population inversion can be maintained with a relatively low pump power. The upper-state lifetime can be measured, for example, by populating the upper laser level with a short laser pulse and monitoring the decay of the fluorescence.

1.8 Allowed and Forbidden Transitions

The fundamental physical process underlying laser operation is the radiative absorption and emission of light by the optically active centers of the laser materials. This involves laser ions undergoing transitions between electronic energy levels through the absorption and emission of photons. The radiative transition strengths or rate for absorption or emission involving specific energy levels are the main parameters of interest. Qualitative estimates of strengths can be determined from group theory considerations. This can be used to classify absorption or emission of radiation between specific energy levels as allowed or forbidden transitions. The quantitative magnitudes of the transition strengths are important for calculating laser gain and threshold conditions.

1.9 Principle of Laser (Absorption and Emission)

From the theory of interaction of radiation with matter, we can get an idea regarding the working of laser. Basically, every laser system essentially has an active or gain medium, placed between a pair of optically parallel and highly reflecting mirrors with one of them partially transmitting, and an energy source to pump active medium. The gain media may be solid, liquid, or gas and have the property to amplify the amplitude of the light wave passing through it by stimulated emission, while pumping may be electrical or optical. The gain medium is used to place between pair of mirrors in such a way that light oscillating between the mirrors passes every time through the gain medium and after attaining considerable amplification, emits through the transmitting mirror.

Consider an atom that has only two energy levels, E_1 and E_2. When it is exposed to radiation having a stream of photons, each with energy $h\upsilon$, leads to the following three distinct processes in the medium.

Basic Processes of Laser :　　　　　　　　　　　　　　(Oct. 2017)

1. **Absorption :** An atom in a lower level absorbs a photon and moves to an upper level. When an atom in the state E_1 absorbs an incident photon of energy $h\upsilon$ (= $E_2 - E_1$) and makes a transition to higher energy state E_2, the process is known as absorption.

2. **Spontaneous emission (April 2018) :** An atom in an upper level can decay spontaneously to the lower level and emits a photon of frequency $h\upsilon$ if the transition between E_2 and E_1 is radiative. This photon has a random direction and phase.

3. **Stimulated emission (April 2018) :** An incident photon causes an upper level atom to decay, emitting a "stimulated" photon whose properties are identical to those of the incident photon. The term "stimulated" underlines the fact that this kind of radiation occurs only if an incident photon is present. The amplification arises due to the similarities between the incident and emitted photons.

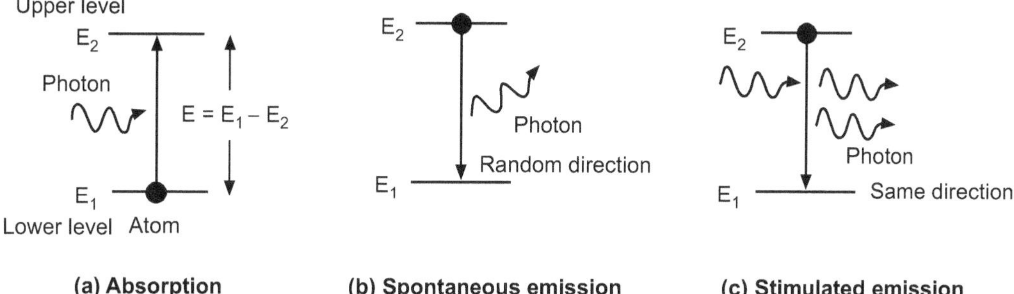

Fig. 1.3 : Mechanism of interaction between atom and photon

The principle of a laser is based on three separate features : (a) Stimulated emission within an amplifying medium, (b) Population inversion of electrons and (c) An optical resonator.

1.10 Einstein's Coefficients and Einstein's Relations

(April 2016, Nov. 2016)

In 1917, Einstein postulated on thermodynamic grounds that the probability for spontaneous emission (A) is related to the probability of stimulated emission (B) and the relationship between them is calculated from quantum mechanics. Einstein's coefficients are mathematical quantities which are a measure of the probability of absorption or emission of light by an atom or molecule.

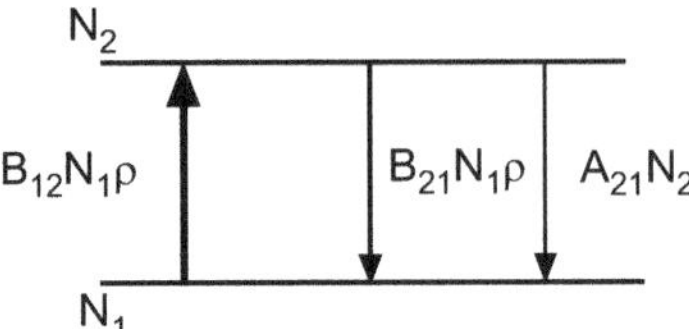

Fig. 1.4

Consider a two level energy system (E_1 and E_2). Let N_1 and N_2 be the number of atoms in the ground state and excited state respectively.

Absorption is the process by which a photon is absorbed by the atom, causing an electron to jump from a lower energy level to a higher one. The process is described by the Einstein's coefficient B_{12}, which gives the probability per unit time per unit spectral energy density of the radiation field that an electron in state 1 with energy E_1 will absorb a photon with an energy $E_2 - E_1 = h\upsilon$ and jump to state 2 with energy E_2. The change in the number density of atoms in state 1 per unit time due to absorption will be :

$$\left(\frac{dN_1}{dt}\right)_{absorb} = -B_{12}N_1\rho(\upsilon) \qquad \dots (1.8)$$

Spontaneous emission is the process by which an electron jumps spontaneously from a higher energy level to a lower one. The process is described by the Einstein's coefficient A_{21} which gives the probability per unit time that an electron in state 2 with energy E_2 will decay spontaneously to state 1 with energy E_1, emitting a photon with an energy $E_2 - E_1 = h\upsilon$. The change in the number density of atoms in state 1 per unit time due to absorption will be :

$$\left(\frac{dN_1}{dt}\right)_{spontaneous} = A_{21}N_2 \qquad \dots (1.9)$$

Stimulated emission is the process by which an electron is induced to jump from a higher energy level to a lower one by the presence of electromagnetic radiation at (or near) the frequency of the transition. From the thermodynamic viewpoint, this process must be regarded as negative absorption. The process is described by the Einstein's coefficient B_{21}, which gives the probability per unit time per unit spectral energy density of the radiation field that an electron in state 2 with energy E_2 will decay to state 1 with energy E_1, emitting a photon with an energy $E_2 - E_1 = h\upsilon$. The change in the number density of atoms in state 1 per unit time due to induced emission will be :

$$\left(\frac{dN_1}{dt}\right)_{stimulated} = B_{21}N_2\rho(\upsilon) \qquad \dots (1.10)$$

At thermodynamic equilibrium, we will have a simple balancing, in which the net change in the number of any excited atoms is zero, being balanced by loss and gain due to all processes.

$$0 = A_{21}N_2 + B_{21}N_2\rho(\upsilon) - B_{12}N_1\rho(\upsilon)$$

$$\rho(\upsilon) = \frac{A_{21}N_2}{B_{12}N_1 - B_{21}N_2}$$

$$\rho(\upsilon) = \frac{\dfrac{A_{21}}{B_{12}}}{\dfrac{N_1}{N_2} - \dfrac{B_{21}}{B_{12}}}$$

From equation (1.7),

$$\rho(\upsilon) = \frac{\dfrac{A_{21}}{B_{12}}}{e^{(E_2 - E_1)/KT} - \dfrac{B_{21}}{B_{12}}}$$

According to Plank's radiation law for any value of T,

$$\frac{A_{21}}{B_{12}} = \frac{8\pi h\upsilon^3 \mu^3}{c^3}$$

$$\rho(\upsilon) = \frac{8\pi h\upsilon^3\mu^3/c^3}{e^{(E_2 - E_1)/KT} - \dfrac{B_{21}}{B_{12}}}$$

where, μ is the refractive index of the medium.

When an atom with two energy levels is placed in the radiation field, then

$$B_{12} = B_{21} \qquad\qquad \text{... (1.11)}$$

Hence, $\qquad \rho(\upsilon) = \dfrac{8\pi h\upsilon^3\mu^3/c^3}{e^{(E_2 - E_1)/KT} - 1}$

and $\qquad \dfrac{A_{21}}{B_{21}} = \dfrac{8\pi h\upsilon^3}{c^3} \qquad\qquad \text{... (1.12)}$

Equations (1.12) and (1.13) are called Einstein's relations between coefficients A_{12} and B_{21} and constants B_{12}, B_{21} and A_{21} are known as Einstein's coefficients.

Solved Examples

Example 1 : Find out the population ratio of the two states that produces a light beam of wavelength 6943 A° in a ruby laser and beam of wavelength 6328 A° in a He-Ne laser at temperature 300°K. (Boltzmann's constant, K = 1.38×10^{-23} J/K) **(April 2017, 2018)**

Solution : Given : λ_1 = 6943 A° = 6943×10^{-10} m, λ_2 = 6328 A° = 6328×10^{-10} m, T = 3000° K, $\dfrac{N_1}{N_2}$ = ?

Formula : The population ratio of the two states is

$$\frac{N_1}{N_2} = e^{-\left(\frac{\Delta E}{KT}\right)}$$

$$\therefore \quad \frac{N_1}{N_2} = e^{-\left(\frac{hc/\Delta\lambda}{KT}\right)}$$

$$\therefore \quad \frac{N_1}{N_2} = \exp\left[\frac{-(6.63 \times 10^{-34} \times 3 \times 10^8)/(6943 - 6328) \times 10^{-10}}{1.38 \times 10^{-23} \times 3000}\right]$$

$$\frac{N_1}{N_2} = \exp\left[\frac{0.03234 \times 10^{-16}}{4140 \times 10^{-23}}\right]$$

$$\frac{N_1}{N_2} = e^{-78.1}$$

$$\therefore \quad \frac{N_1}{N_2} = 1.2 \times 10^{-34}$$

Hence, the population ratio of the two states that produces a light beam is 1.2×10^{-34}.

Example 2 : Calculate the wavelength, frequency and energy per pulse of CO_2 beam having energy difference between two states as 0.117 eV. The laser contains total of 2.5×10^{19} atoms of elements in excited state.

Solution : Given : ΔE = 0.117 eV = $0.117 \times 1.6 \times 10^{-19}$ J, N = 2.5×10^{19} atoms, λ = ?, υ = ? and energy per pulse = ?

Formula : The Energy relation is,

$$E = \frac{hc}{\lambda}$$

(a) Wavelength of laser is,

$$\lambda = \frac{hc}{E}$$

$$= \frac{6.63 \times 10^{-34} \times 3 \times 10^8}{0.117 \times 1.6 \times 10^{-19}}$$

$$\lambda = 106.25 \times 10^{-7} \text{ m } \textbf{or } 10.625 \text{ } \mu m$$

(b) Frequency of laser is,

$$E = h\upsilon$$

$$\upsilon = \frac{E}{h}$$

$$\upsilon = \frac{0.117 \times 1.6 \times 10^{-19}}{6.63 \times 10^{-34}} = 0.02823 \times 10^{15}$$

$$\upsilon = 2.823 \times 10^{13} \text{ Hz}$$

(c)　　　Energy per pulse = Energy $\times$ Total number of excited atoms

$$= 0.117 \times 1.6 \times 10^{-19} \times 2.5 \times 10^{19}$$

$$= 0.468 \text{ J}$$

Example 3 : Find out the atomic population at room temperature at the first excited level for hydrogen gas having excited energy state 3.39 eV.

Solution : Given : $E_2 = 3.39$ eV $= 3.39 \times 1.6 \times 10^{-19}$ J, $E_1 = 13.6$ eV $= 13.6 \times 1.6 \times 10^{-19}$ J,

$$T = 27°C = 300°K, \quad \frac{n_1}{n_2} = ?$$

Formula : The population ratio of the two states is

$$\frac{N_1}{N_2} = e^{\left(\frac{\Delta E}{KT}\right)}$$

$$= \exp\left[\frac{(3.39 - 13.6) \times 1.6 \times 10^{-19}}{1.38 \times 10^{-23} \times 300}\right]$$

$$= \exp\left[\frac{-16.336 \times 10^{-19}}{414 \times 10^{-23}}\right]$$

$$= e^{-394.6}$$

$$\therefore \qquad \frac{N_1}{N_2} = 0$$

Hence population inversion is not possible.

Example 4 : Find out the atomic population at temperature 6000 K at the first excited level for hydrogen gas having excited energy state 3.39 eV.

Solution : Given : $E_2 = 3.39$ eV $= 3.39 \times 1.6 \times 10^{-19}$ J, $E_1 = 13.6$ eV $= 13.6 \times 1.6 \times 10^{-19}$ J,

$$T = 6000°K, \quad \frac{N_1}{N_2} = ?$$

Formula : The population ratio of the two states is

$$\frac{N_1}{N_2} = e^{\left(\frac{\Delta E}{KT}\right)}$$

$$\frac{N_1}{N_2} = \exp\left[\frac{(3.39 - 13.6) \times 1.6 \times 10^{-19}}{1.38 \times 10^{-23} \times 6000}\right]$$

$$\frac{N_1}{N_2} = \exp\left[\frac{-16.336 \times 10^{-19}}{8.28 \times 10^{-20}}\right]$$

$$= e^{-19.73}$$

$$\therefore \qquad \frac{N_1}{N_2} = 2.5 \times 10^{-9}$$

Example 5 : Calculate the coefficient for stimulated emission if the wavelength of emission radiation is 6300 A°. Time is 1 μsec.

Solution : Given : $\lambda = 6300\ A^\circ = 6300 \times 10^{-10}$ m , $t_{sp} = 1\ \mu sec = 1 \times 10^{-6}$ sec, $A_{21} = ?$

Formula : Einstein's coefficient relation is

$$\frac{A_{21}}{B_{21}} = \frac{8\pi h \mu^3 \upsilon^3}{c^3}$$

But, $A_{21} = \dfrac{1}{t_{sp}}$ and $\dfrac{c^3}{\upsilon^3} = \lambda^3$,

$$\frac{1}{t_{sp}\, B_{21}} = \frac{8\pi h \mu^3 \upsilon^3}{c^3} = \frac{8\pi h \mu^3}{\lambda^3}$$

$$B_{21} = \frac{\lambda^3}{8\pi h \mu^3 t_{sp}}$$

$$B_{21} = \frac{(6300 \times 10^{-10})^3}{8\pi \times 6.63 \times 10^{-34} \times 1^3 \times 1 \times 10^{-6}}$$

$$B_{21} = 1.5 \times 10^{19}\ m^3\ J^{-1}\ s^{-2}$$

Hence, the coefficient for stimulated emission is $1.5 \times 10^{19}\ m^3\ J^{-1}\ s^{-2}$

Exercise

(A) Multiple Choice Questions :

1. Principle of laser is ......

   (a) spontaneous absorption   (b) simulated emission

   (c) both b and c

2. Planck's constant = ...

   (a) 6.62×10^{-34} J.sec   (b) 6.62×10^{-34} J.min

   (c) 6.62×10^{-34} cal.sec

3. The ordinary light consists of wavelengths roughly in the range of ...... nanometers.

(a) 200-700   (b) 300-700

(c) 400-700

4. The typical wavelength of a ruby laser is ......

(a) 6943 A°   (b) 6328 A°

(c) 6600 A°

5. The typical wavelength of a He-Ne laser is ......

(a) 6943 A°   (b) 6328 A°

(c) 6600 A°

6. Who investigated first MASER device and when?

(a) Charles H. Townes, 1951   (b) Ali Javan, 1960

(c) Charles H. Townes, 1964

7. Who investigated first LASER and when?

(a) Charles H. Townes, 1951   (b) Ali Javan, 1960

(c) Charles H. Townes, 1964

8. Two light-sources are said to be coherent if they vibrate ....

(a) in same phase   (b)   with constant phase difference

(c) both a and b

Answers : (1) b, (2) a, (3) c, (4) a, (5) b, (6) a, (7) b, (8) c.

(B) Short Answer Questions :

1. What is laser light?

2. Who and when investigated first MASER device?

3. Who and when investigated first LASER?

4. Draw a diagram for stimulated emission of radiation in laser mechanism.

5. State the relation between Einstein's coefficients.

6. Define population density.

7. Define energy levels.

8. State Boltzmann's equation at thermal equilibrium.

(C) Long Answer Questions :

1. What is laser light? Distinguish between ordinary light and laser light.
2. Explain in brief history of Laser light.
3. Obtain Einstein's coefficient relations in laser action.
4. Explain basic three processes with neat diagram in brief.
5. Describe importance of energy level diagram in laser.
6. Explain interaction of light with matter.

Chapter **2**...

Laser Action

Objectives ...

- To study and derive the condition for large stimulated emission.
- To explore the different methods of population inversion.
- To study and derive condition for light amplification.
- To understand the fundamental terms such as Gain coefficient, Active medium, Metastable states.
- To understand and study the different fundamental pumping schemes such as three-level and four-level pumping schemes.

2.1 Condition for Large Stimulated Emission

Lasing is the process which leads the emission of stimulated photons due to the transition of atoms from the metastable state to the ground state after achieving population inversion. For the generation of laser, stimulated emission is essential. Stimulated emission has to be induced or stimulated and is generated under special conditions as stated by Einstein. When the population inversion exists between upper and lower levels among atomic systems, it is possible to realize amplified stimulated emission and the stimulated emission has the same frequency and phase as the incident radiation.

The absorption and spontaneous emission always occurs together with stimulated emission.

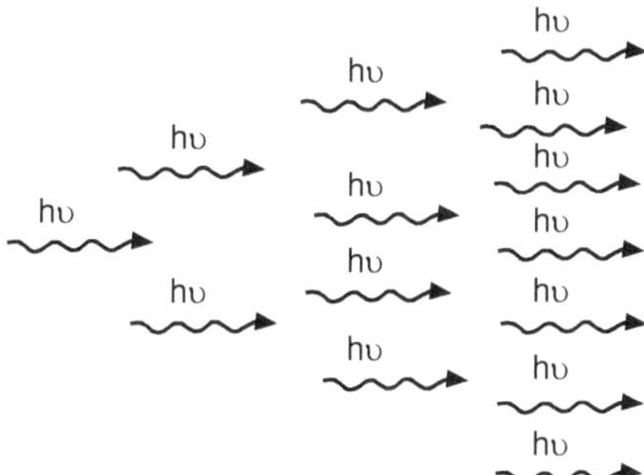

Fig. 2.1 : Stimulated emission

As we know that, Einstein's coefficient relation is

$$\frac{A_{21}}{B_{12}} = \frac{8\pi h\upsilon^3}{c^3} \qquad \qquad \dots (2.1)$$

Therefore, the ratio of stimulated transitions to spontaneous transitions is

$$R = \frac{\text{Stimulated transitions}}{\text{Spontaneous transitions}}$$

$$R = \frac{\left(\dfrac{dn_1}{dt}\right)_{\text{stimulated}}}{\left(\dfrac{dn_1}{dt}\right)_{\text{spontaneous}}} = \frac{B_{21}n_2\rho(\upsilon)}{A_{21}n_2}$$

$$R = \frac{B_{21}}{A_{21}}\rho(\upsilon) \qquad \qquad \ldots (2.2)$$

But we know that,

$$\rho(\upsilon) = \frac{8\pi h\upsilon^3\mu^3/c^3}{e^{(E_2 - E_1)/KT} - 1}$$

Therefore, equation (2.2) becomes

$$R = \left(\frac{B_{21}}{A_{21}}\right)\frac{8\pi h\upsilon^3\mu^3/c^3}{e^{(E_2 - E_1)/KT} - 1}$$

From equation (2.1), $\qquad R = \left(\dfrac{1}{8\pi h\upsilon^3\mu^3/c^3}\right)\dfrac{8\pi h\upsilon^3\mu^3/c^3}{e^{(E_2 - E_1)/KT} - 1}$

$$R = \frac{1}{e^{(E_2 - E_1)/KT} - 1} \qquad \qquad \ldots (2.3)$$

Equation (2.3) gives ratio of stimulated transitions to spontaneous transitions if both are equal.

i.e. $\qquad\qquad\qquad R = \dfrac{1}{2}$

At thermodynamic equilibrium,

$$B_{12} = B_{21}$$

Then, $\qquad\qquad\qquad R = \dfrac{n_2}{n_1} \qquad \qquad \ldots (2.4)$

and $\qquad\qquad\qquad \dfrac{n_2}{n_1} << 1$

The two equations (2.3) and (2.4) are to be satisfied to make stimulated emission overwhelm the spontaneous emission.

2.2 Population Inversion　　　　　　　　　　**(April 2018)**

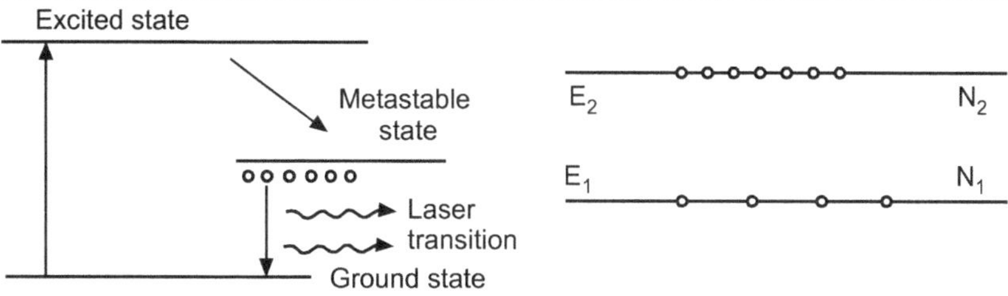

Fig. 2.2 : Population inversion

Photon absorption and emission process takes place in an atomic system at thermal equilibrium. However, laser operation requires obtaining stimulated emission. The non-equilibrium state, where the population of upper energy level N_2 is more than the population of lower energy level N_1, is known as state of population inversion. State of population inversion is referred to as negative temperature state, means, state of population inversion is non-equilibrium state. This implies population inversion state is attained at normal temperature. Population inversion is a state of the system which deviates from thermal equilibrium. In thermal equilibrium, the population of the lower level is always higher and a positive net gain can never occur. Population inversion is sometimes described as a state with a negative temperature. In many cases, it is achieved by optical pumping. Population inversion is an artificial situation that is established by generating a large number of atoms in the higher energy state than that of the lower energy state i.e. $n_2 >> n_1$. This is known as **population inversion condition**.　　　　　**(April 2016)**

There are several methods for achieving the condition of population inversion necessary for the laser action to take place. The most common methods used are optical pumping and direct electron excitation, inelastic atom-atom collisions and chemical reaction.

2.3 Condition for Light Amplification

Let us consider a beam of light propagating through active material medium. When the photons are incident on unexcited atoms, they lose energy. This loss of energy may be absorbed by atoms of material medium and those atoms jump to the excited state. An incident photon may cause stimulated emission in one of these atoms. It produces two or more coherent photons of same frequency. These photons produce four more coherent photons. All these photons moving in same direction are added to light beam and increase its intensity. This process is called light amplification.

Let, $\dfrac{-dN}{dt}$ be the rate of loss of photons from beam when they travel through medium having thickness x and area of unity.

Therefore, the rate of absorption and stimulated emission is

$$\frac{-dN}{dt} = B_{12}\, \rho(\upsilon)\, N_1 - B_{21}\, \rho(\upsilon)\, N_2$$

If the coefficient of absorption and stimulated emission is equal ($B_{12} = B_{21}$), then

$$\frac{-\,dN}{dt} = B_{12}\,\rho(\upsilon)\,(N_1 - N_2) \qquad \ldots (2.5)$$

The intensity of light field is equal to

$$I = \rho(\upsilon)\,\upsilon = nh\nu\upsilon \qquad (\text{since, } \rho(\upsilon) = nh\nu) \ldots (2.6)$$

The loss of photons ($-\,dN$) in a small thickness dx of medium is

$$-\,dN = \frac{dI}{dx}\frac{dx}{h\upsilon}$$

$$\frac{-\,dN}{dt} = \frac{dI}{dx}\frac{dx}{dt}\frac{dx}{h\upsilon}$$

$$\frac{-\,dN}{dt} = \frac{dI}{dx}\left(\frac{1}{h\upsilon}\right) \qquad \left(\text{since } \frac{dx}{dt} = v\right) \ldots (2.7)$$

But,

$$\frac{dI}{dx} = \alpha\,\rho(\upsilon)\,\upsilon$$

Therefore,

$$\frac{-\,dN}{dt} = \frac{\alpha\,\rho(\upsilon)\,\upsilon}{h\upsilon} \qquad \ldots (2.8)$$

Comparing equations (2.5) and (2.8) we get

$$\frac{\alpha\,\rho(\upsilon)\,\upsilon}{h\upsilon} = B_{12}\,\rho(\upsilon)\,(N_1 - N_2)$$

$$\alpha = \frac{B_{12}\,(N_1 - N_2)}{\upsilon}\,(h\upsilon) \qquad \ldots (2.9)$$

Above equation (2.9) implies absorption coefficient α is related to difference in population of two energy levels i.e. $\alpha \propto (N_1 - N_2)$. For $(N_2 - N_1)$, α becomes negative. Then

$$\alpha = \frac{B_{12}\,(N_2 - N_1)}{\upsilon}\,(h\upsilon) \qquad \ldots (2.10)$$

This is the condition for light amplification, where, α is called gain coefficient per length.

If $N_1 > N_2$, then α is positive and if $N_2 > N_1$, then α is negative.

Positive α implies that intensity of light increases exponentially when beam travels through medium. We rewrite this equation by putting the value of B_{21}.

$$\therefore \qquad \alpha = (N_2 - N_1)\frac{v^2}{8\pi v^2 \tau} \qquad \ldots (2.11)$$

2.4 Active Medium and Metastable States (April 2018, 2017; Oct. 2017)

Active medium (April 2017) : A medium in which light gets amplified is called an active medium. The medium may be solid, liquid or gas. Out of the different atoms in the medium only small fraction of the atoms is responsible for stimulated emission and consequent light amplification.

Metastable state (April 2016, Nov. 2016) : The higher energy state of an atom (or molecule) in which it stays for unusually longer duration of time (of the order of ms or more) is called metastable state

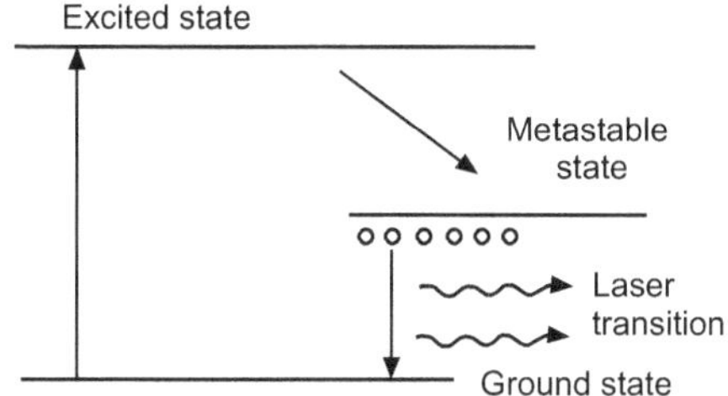

Fig. 2.3 : Understanding of metastable state

2.5 Pumping Schemes : Three-Level and Four-Level (Nov. 2016)

Pumping is the phenomenon of achieving population inversion, i.e the process which raises the atoms from lower energy state to higher energy state in the active medium.

(April 2017, 2018)

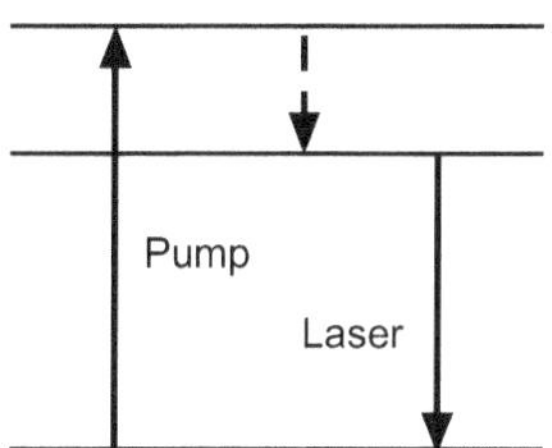

Fig. 2.4 : Understanding of pumping of atoms

The various techniques to pump the atom in excited state are : (Oct. 2017)

1. **Optical Pumping :** A light source is used to supply luminous energy and create population inversion by optical photon. For example : Ruby laser.

2. **Electrical Pumping :** Electrical discharge converts the gas medium into plasma which liberates electrons which, in turn, are accelerated by the strong electric fields present in the tube. These electrons, on collision with neutral gas atoms, make some atoms to jump to excited state. For example : He-Ne laser

3. **Chemical Pumping :** An exothermic chemical reaction is used to produce energy. In chemical lasers energy comes from chemical reactions without any need for other energy sources.

4. **Direct Conversion :** Electrical energy is directly converted into radiation in devices like LEDs and semiconductor lasers.

Pumping scheme is classified as two-level, three-level and four-level optical pumping schemes.

(a) Three-level optical pumping scheme :

Three-level pumping scheme is one, in which the lower laser level is either the ground state or level whose separation from the ground state is small compared to kT. Initially the population distribution among the three levels obeys the Boltzmann's law .When the atoms are subjected to an intense radiation of pumping frequency, $\upsilon_p = (E_3 - E_1/h)$, the atoms are pumped to higher level E_3.

Some of the excited atoms make spontaneous transition to the ground state, but many of them undergo spontaneous non-radioactive transitions to the metastable level E_2. As spontaneous transitions from E_2 to E_1 do not occur often, the atoms are accumulated at the metastable level E_2. The build-up of atoms at E_2 level becomes continuous because of pumping process. Eventually, the population N_2 at E_2 exceeds the population N_1 at E_1 and population inversion is attained. Now a photon of energy $h\upsilon = E_2 - E_1$ can induce stimulated emission and laser action.

In these schemes the terminal level of the laser transition is simultaneously the ground state. Therefore, inversion of population requires more than half of the ground state atoms to be lifted to the higher energy level. As the ground state is heavily populated, large pumping power is to be used to depopulate the ground level to the required extent.

The three-level pumping scheme can produce light only in pulses. Once stimulated emission commences, the metastable state E_2 gets depopulated very rapidly and the population of the ground state increases quickly. As a result, the population inversion is ended.

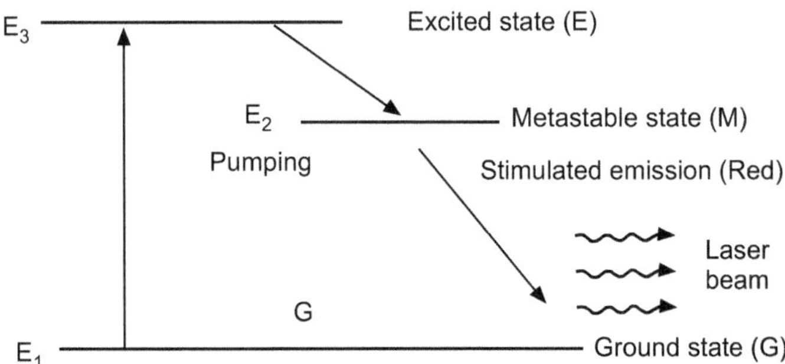

Fig. 2.5 : Three-level optical pumping

(b) Four-level optical pumping scheme :   (April 2016, Nov. 2016, Oct. 2017)

In this scheme the terminal laser level E_2 is well above the ground level such that $(E_2 - E_1) >> KT$. Therefore, the thermal equilibrium population of E_2 level is negligible. As in the three-level pumping schemes, the pump energy elevates the atoms to a short lived uppermost level E_4. The atoms then drop spontaneously to a metastable upper laser level E_3

as the terminal laser level E_2 is virtually vacant, population inversion between the states E_3 and E_2 is quickly established. A spontaneous photon of energy $h\upsilon = E_3 - E_2$ can initiate a chain of stimulated emissions culminating in lasing. The laser transition takes the atoms to the level E_2. From there the atoms lose the rest of their excess energy by radiative or non-radiative transitions and finally reach the grounded state E_1.

The lower laser transition level in four-level scheme is not the ground state and is virtually vacant. As soon as some atoms are pumped to the upper laser level, population inversion is achieved. Thus, it requires less pumping energy than those in a three-level laser. This is the major advantage of this scheme. Four-level lasers can operate in a continuous wave (CW) mode.

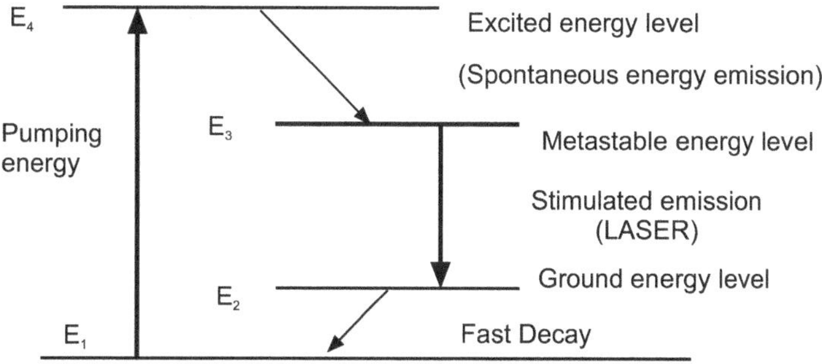

Fig. 2.6 : Four-level optical pumping

Solved Examples

Example 1 : Calculate the temperature at which rate of spontaneous emission is equal to rate stimulated emission for the wavelength of beam 5500 A°.

Solution : Given : $\lambda = 5500$ A° $= 5500 \times 10^{-10}$ m, $R = \dfrac{1}{2}$, T = ?

Formula : Rate of spontaneous emission is equal to rate stimulated emission.

$$R = \frac{1}{e^{\Delta E/KT} - 1}$$

$$\frac{1}{2} = \frac{1}{e^{\Delta E/KT} - 1}$$

$$2 = e^{\Delta E/KT} - 1$$

Taking natural log on both sides, we get

$$\ln 2 = \frac{\Delta E}{KT}$$

$$0.693 = \frac{hc}{\lambda KT}$$

Therefore,

$$T = \frac{hc}{0.693\, K\lambda}$$

$$T = \frac{6.63 \times 10^{-34} \times 3 \times 10^{8}}{0.693 \times 1.38 \times 10^{-23} \times 5500 \times 10^{-10}}$$

$$T = 37813 \,^{\circ}K$$

Example 2 : Calculate the wavelength of beam at which rate of spontaneous emission is equal to rate stimulated emission for the temperature of 300 °K.　　**(April 2016, Oct. 2017)**

Solution : Given : $R = \dfrac{1}{2}$, T = 300 °K, λ = ?

Formula : Rate of spontaneous emission is equal to rate stimulated emission.

$$R = \frac{1}{e^{\Delta E/KT} - 1}$$

$$\frac{1}{2} = \frac{1}{e^{\Delta E/KT} - 1}$$

$$2 = e^{\Delta E/KT} - 1$$

Taking natural log on both sides, we get

$$\ln 2 = \frac{\Delta E}{KT}$$

$$0.693 = \frac{hc}{\lambda KT}$$

$$\therefore \quad \lambda = \frac{hc}{0.693\, KT}$$

$$\lambda = \frac{6.63 \times 10^{-34} \times 3 \times 10^{8}}{0.693 \times 1.38 \times 10^{-23} \times 300}$$

$$\lambda = 69.3 \times 10^{-6} \text{ m or } 69.3 \,\mu m$$

Exercise

(A) Multiple Choice Questions :

1. Pumping source preferred for gaseous lasers is ......

   (a) optical pumping　　　(b) electrical pumping　　(c) chemical pumping

2. Minimum value of energy in eV required to excite an electron from ground state to higher state corresponds to ......

   (a) excitation potential　　　(b) ionization potential　　(c) critical potential

3. Can we obtain light amplification in absence of stimulated emission?

   (a) Yes   (b) No

4. In thermal equilibrium, during lasing action, the upward and downward transitions between ground state and excited state are ....

   (a) equal   (b) not equal   (c) none

5. Stimulated emission depends on ....

   (a) the number of atoms present in the excited state

   (b) the intensity of incident light

   (c) both

6. In lasing action, the spontaneous emission does not depend on......

   (a) the number of atoms present in the excited state

   (b) the intensity of incident light

   (c) both

7. The population inversion process is observed due to the existence of ......

   (a) metastable state   (b) excited state   (c) ground state

8. The concept of stimulated emission was first put forward by ......

   (a) Einstein   (b) Newton   (c) Openheimer

9. The life time of metastable state in comparison to excited state is ......

   (a) smaller   (b) greater   (c) equal

10. In the absorption process, the photon is ....

   (a) lost   (b) created   (c) none

Answers : (1) a, (2) a, (3) b, (4) a, (5) c, (6) b, (7) a, (8) a, (9) a, (10) a.

(B) Short Answer Questions :

1. Distinguish between spontaneous and stimulated emissions.

2. What is the requirement to produce laser beam?

3. What are the various techniques of pumping?

4. What do you mean by population inversion?

5. What is Pumping?

6. Define metastable state in laser.

7. Define active medium in laser.

8. State condition for population inversion.

(C) Long Answer Questions :

1. What is pumping? Explain three-level optical pumping in laser.

2. What is pumping? Explain four-level optical pumping in laser.

3. What is Pumping? Explain the various techniques of pumping.

4. Explain how stimulated emission overwhelm the spontaneous emission in condition for large stimulated emission.

5. Obtain the condition for light amplification in laser action.

Chapter **3**...

Laser Oscillator

Objectives ...

- To understand optical feedback in laser.
- To study and derive expression for threshold gain.
- To study and obtain expression for critical population inversion.
- To understand the term optical resonator.
- To obtain fundamental condition for steady state oscillations.
- To derive expression for cavity resonance frequencies.

3.1 Optical Feedback (April 2016, 2017, 2018; Oct. 2017)

The laser is an optical oscillator. It comprises a resonant optical amplifier whose output is fed back into its input with matching phase as shown in figure.

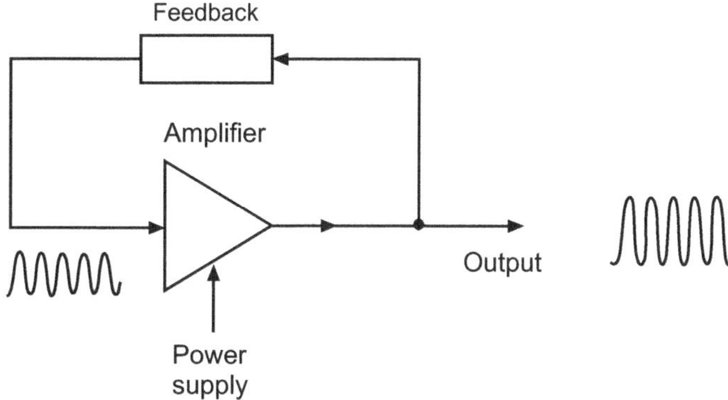

Fig. 3.1 : Feedback oscillator

In the absence of such an input there is no output, so that the feedback signal is also zero. However, this is an unstable situation. The presence of even a small amount of noise at the input is unavoidable and may initiate the oscillation process. The input is amplified and the output is fed back to the input, where it undergoes further amplification. The process continues indefinitely until a large output is produced. Saturation of the amplifier gain limits further growth of the signal, and the system reaches a steady state in which an output signal is created at the frequency of the resonant amplifier.

When the oscillator is switched on, any noise signal of the approximate frequency appearing at the input will be amplified. The amplified output gives feedback to the input. A stable output is quickly reached and the oscillator acts as a source of the particular frequency.

(3.1)

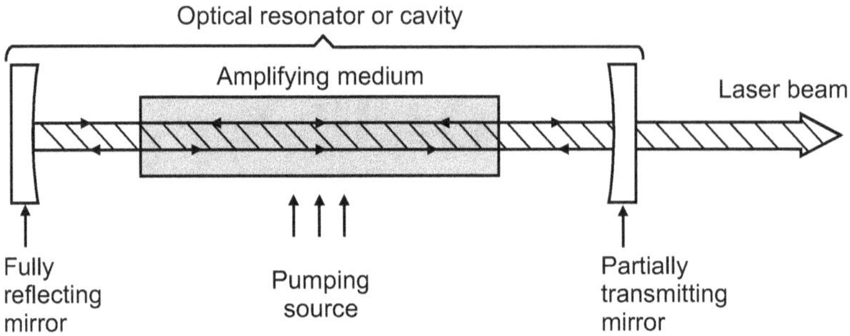

Fig. 3.2 : Optical oscillator

The laser is an optical oscillator in which the amplifier is the pumped active medium. Gain saturation is a basic property of laser amplifiers. Feedback is obtained by placing the active medium in an optical resonator, which reflects the light back and forth between its mirrors. Frequency selection is achieved by the resonant amplifier and by the resonator, which admits only certain modes. Output coupling is accomplished by making one of the resonator mirrors partially transmitting.

Laser Amplification (gain coefficient) :

The laser amplifier is a narrowband coherent amplifier of light. Amplification is achieved by stimulated emission from an atomic or molecular system with a transition whose population is inverted i.e., the upper energy level is more populated than the lower. The amplifier bandwidth is determined by the linewidth of the atomic transition, or by an inhomogeneous broadening mechanism such as the Doppler Effect in gas lasers. The laser amplifier is a distributed-gain device characterized by its gain coefficient (gain per unit length) $\gamma(\upsilon)$, which governs the rate at which the photon-flux density φ increases. When the photon-flux density φ is small, the gain coefficient is

$$\gamma_0(\upsilon) \;=\; N_0\sigma(\upsilon) = N_0\,\frac{\lambda^2}{8\pi t_{sp}}\,g(\upsilon) \qquad\qquad \dots (3.1)$$

where, N_0 = equilibrium population density difference (density of atoms in the upper energy state minus that in the lower state); N, increases with increasing pumping rate

$$\sigma(\upsilon) \;=\; (\lambda^2/8\pi t_{sp})\,g(\upsilon) = \text{transition cross-section}$$

$$t_{sp} \;=\; \text{spontaneous lifetime}$$

$$g(\upsilon) \;=\; \text{transition lineshape}$$

$$\lambda \;=\; \text{wavelength in the medium} = \lambda_0/n$$

$$n \;=\; \text{refractive index.}$$

3.2 Round Trip Gain　　　　　(April 2016, 2017; Nov. 2016)

In laser, gain is a process where the medium transfers part of its energy to the emitted electromagnetic radiation, resulting an increase in optical power. This is the basic principle of all lasers. Quantitatively, gain is a measure of the ability of a laser medium to increase optical power.

Round-trip gain means gain multiplied by the length of propagation of the laser emission during a single round-trip. In the case of gain varying along the length, the round-trip gain can be expressed with integral;

$$g = \int G\, dz \qquad \qquad \text{... (3.2)}$$

where, G is laser gain (amplification) and z is the co-ordinate in the direction of propagation.

3.3 Threshold Gain

The lasing threshold is the lowest excitation level at which output of laser is dominated by stimulated emission rather than by spontaneous emission. Below the threshold, the output power of laser rises slowly with increasing excitation. Above threshold, the slope of power vs. excitation is orders of magnitude greater. The linewidth of the laser emission also becomes the orders of magnitude smaller above the threshold than it is below. Above the threshold, the laser is said to be lasing. In optical resonator, light undergoes amplification and suffers various losses. These losses occur due to transmission at the output mirror, scattering and diffraction of light within the active medium. For the proper build up of oscillations, it is essential that the amplification between two consecutive reflections of light from rear and mirror can balance the losses. We can determine the threshold gain by considering the change in intensity of a beam of light undergoing a round trip within the resonator.

Let us assume that the laser medium fills the space between the mirrors M_1 and M_2 which has reflectivity r_1 and r_2 respectively. Let the mirrors may be separated by a distance L. Further, let the intensity of the light beam be I_0 at M_1. Then in travelling from mirror M_1 to mirror M_2, the beam intensity increases from I_0 to I_L, which is given by,

$$I_L = I_0\, e^{(\gamma - \alpha_s)\, L} \qquad \qquad \text{... (3.3)}$$

After reflection at M, the beam intensity will be $r_2\, I_0\, e^{(\gamma - \alpha_s)\, L}$ and after a complete round trip the final intensity will be,

$$I_{2L} = r_1\, r_2\, I_0\, e^{(\gamma - \alpha_s)\, 2L} \qquad \qquad \text{... (3.4)}$$

The amplification obtained during the round trip is,

$$G = \frac{I\,(2L)}{I_0} = r_1\, r_2\, I_0\, e^{(\gamma - \alpha_s)\, 2L}$$

The product '$r_1 r_2$' represents the losses at the mirrors, whereas 'α_s' includes all the distributed losses such as scattering, diffraction and absorption occurring in the medium.

The losses are balanced by gain, when $G \geq 1$ or $I_{2L} = I_0$.

It leads to the condition that,

$$r_1\, r_2\, I_0\, e^{(\gamma - \alpha_s)\, 2L} \geq 1$$

$$e^{(\gamma - \alpha_s)\, 2L} \geq \frac{1}{r_1 r_2}$$

Taking logarithm on both sides, we get,

$$2L\,(\gamma - \alpha_s) \geq -\ln(r_1 r_2)$$

$$(\gamma - \alpha_s) \geq \frac{1}{2L}\ln(r_1 r_2)$$

$$\gamma \geq \alpha_s - \ln\frac{1}{2L} r_1 r_2$$

$$\gamma \geq \alpha_s + \frac{1}{2L}\ln\frac{1}{r_1 r_2} \qquad \ldots (3.5)$$

Equation (3.5) is known as the Condition for Lasing. It implies that the initial gain must exceed the sum of the losses in the cavity and the amplification of the laser 'γ' will be dependent on how hard the laser medium is pumped. This condition is used to determine the threshold value of pumping energy necessary for lasing action.

3.4 Critical Population Inversion　　　　　(Nov. 2016, Oct. 2017)

Critical population inversion is the minimum population inversion density required to start lasing action and then to sustain it. Therefore, critical population inversion for quantity N_{th} is

$$N_{th} = (N_2 - N_1)_{th} \qquad \ldots (3.6)$$

In order to achieve critical population condition with lowest pumping power, atom must have narrow line width i.e. $\Delta\upsilon$. The lasing condition under critical population inversion is

$$N_{th} = [(4\pi\upsilon_0\, 2\tau_{sp}\, \gamma_{th}\, \Delta\upsilon)/\upsilon^2]/(l_c/L) \qquad \ldots (3.7)$$

where, υ_0 is central frequency, l_c is fractional loss, $\Delta\upsilon$ is linewidth, L is length of cavity, γ_{th} is the amplification, τ_{sp} is rate of spontaneous emission and υ is frequency of beam.

3.5 Optical Resonator　　　　　(April 2017, Nov. 2016)

Optical resonator is an arrangement of mirrors that forms a standing wave cavity resonator for light waves. Optical cavities are a major component of lasers, surrounding the gain medium and providing feedback of the laser light. Optical feedback is achieved by placing the active medium in an optical resonator. A Fabry-Perot resonator, comprising two mirrors separated by a distance L, contains the medium (refractive index n) in which the active atoms of the amplifier reside.

The large number of excited atoms produced due to population inversion and pumping emit photons spontaneously in various directions. These photons, in turn, strike atoms in metastable state and cause stimulated emission. The photons produced due to stimulated emission also travel in various directions. Since these photons cannot give coherent beam, the number of photon states must be restricted to obtain a coherent beam of laser.

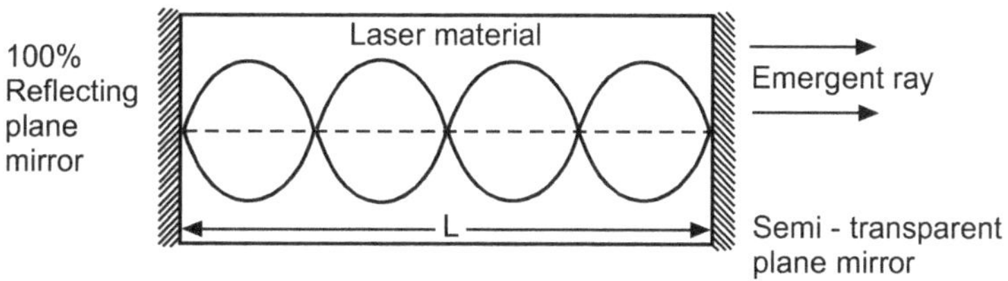

Fig. 3.3　　　　　　　　　　　　**(April 2017)**

This can be achieved by placing the active medium between a perfectly reflecting plane spherical mirror and a semi-transparent mirror (90% reflecting). The mirror system reflects most of the energy from the light incident on it back to the medium, thus acting as a positive feedback necessary to compensate the losses. A small amount of energy escapes through the semi-transparent mirror to form the laser. If enough population inversion takes place in the active medium, the light is amplified to a great extent since each passage of light across the medium after successive reflections causes gain in strength and multiple reflections occur. A steady intense laser beam emerges from the semi-transparent medium.

3.6 Condition for Steady State Oscillations

Light amplification is a continuous increase in amplitude. It is necessary that wave making a complete round trip inside the resonator has to fulfill a phase condition in addition to amplitude condition. The energy of radiation is;

$$E = h\upsilon \qquad \qquad \ldots (3.8)$$

Wave returning to some points in the medium must have same phase as that of original wave with any number of reflections from the mirror is necessary for lasing action. This implies that phase delay between waves is multiple of 2π. Thus, light starts at a wave peak when it is reflected from output mirror. It should be again at the wave peak after one round trip. This implies constant relation between constant λ and length of laser rod L. It is required that optical path length travelled by a wave between two reflections at the same end of mirror should be integral multiple of the wavelength. Therefore, it achieves the condition for amplification.

i.e. $\qquad \qquad 2\mu L = m\lambda \quad$ where $(m = 1, 2, 3 \ldots)$

$$L = \frac{m\lambda}{2\mu} \qquad \qquad \ldots (3.9)$$

where, μ is refractive index of active medium, μL is optical path and L is length of cavity.

This equation expresses the condition of resonance between mirror cavity and light waves.

3.7 Cavity Resonance Frequencies

It is a frequency of radiation in cavity that resonates with the radiation waves which is fit an integral number of half waves between mirrors.

The wavelength of such wave is,

$$2\mu L = m\lambda$$

$$\therefore \qquad \lambda = \frac{2\mu L}{m} \qquad\qquad \dots (3.10)$$

But frequency in terms of wavelength is,

$$\upsilon = \frac{c}{\lambda}$$

Therefore, Cavity resonance frequency is,

$$\upsilon = \frac{mc}{2\mu L} \qquad\qquad (3.11)$$

Solved Examples

Example 1 : Calculate gain coefficient for ruby laser generating wavelength 6943 A° which uses active material having refractive index 1.78 and population inversion density $N_o = 5 \times 10^{17}$ per cm^3. **(April 2017)**

Assume $g(\upsilon) = 5 \times 10^{-12}$ sec and $t_{sp} = 3$ ms.

Solution : Given : $\lambda = 6943$ A° $= 6943 \times 10^{-8}$ m, $\mu = 1.78$, $g(\upsilon) = 5 \times 10^{-12}$ sec

$N_o = 5 \times 10^{17}$ per cm^3, $t_{sp} = 3$ msec $= 3 \times 10^{-3}$ sec, $\gamma_0(\upsilon) = ?$

Formula : Gain coefficient for laser is,

$$\gamma_0(\upsilon) = N_o \frac{\lambda^2}{8\pi\, t_{sp}} g(\upsilon)$$

$$\gamma_0(\upsilon) = 5 \times 10^{17} \times \frac{(6943 \times 10^{-8})^2}{8\pi \times 3 \times 10^{-3}} \times 5 \times 10^{-12}$$

$$\gamma_0(\upsilon) = 0.159$$

Example 2 : Calculate length of cavity and optical path for He-Ne laser tube which achieves the condition for amplification having wavelength 6328 A° and refractive index of active medium between two neighbouring reflectors is 1.56. Also calculate resonant frequency. **(April 2017)**

Solution : Given : $\lambda = 6328$ A° $= 6328 \times 10^{-10}$ m, $\mu = 1.56$, $L = ?$, $\upsilon = ?$, $m = 1$,

Optical path $= ?$

Formula : (a) Length of cavity for laser is,

$$L = \frac{m\lambda}{2\mu} = \frac{1 \times 6328 \times 10^{-10}}{2 \times 1.56} = 2028.2 \times 10^{-10} \text{ m}$$

Hence, Length of cavity for laser is 2028.2×10^{-10} m $= 2028.2$ A°.

(b) Optical path for laser is,

$$\mu L = \frac{m\lambda}{2} = \frac{1 \times 6328 \times 10^{-10}}{2}$$

$$\mu L = 3164 \times 10^{-10} \text{ m}$$

Hence, optical path for laser is 3164×10^{-10} m $= 3164$ A°.

(c) Resonant frequency is,

$$\upsilon = \frac{mc}{2\mu L} = \frac{1 \times 3 \times 10^8}{2 \times 1.56 \times 2028.2 \times 10^{-10}}$$

$$\upsilon = 0.4741 \times 10^{15} \text{ Hz} = 47.41 \times 10^{13} \text{ Hz}$$

Hence, Resonant frequency for laser is 47.41×10^{13} Hz.

Example 3 : What will be the reflectivity of other cavity mirror if the reflectance of first mirror is 100% ? The length of the cavity is 15 cm and gain factor of laser material is 0.0005 per cm. **(April 2016, 2018; Nov. 2016)**

Solution : Given : r_1 = 100% = 1, L =15 cm, γ = 0.0005 per cm, r_2 = ?

Formula : Gain factor of laser material is,

$$\gamma = \frac{1}{2L} \ln \frac{1}{r_1 r_2}$$

$$2L\,\gamma = \ln \frac{1}{r_1 r_2}$$

$$e^{2L\gamma} = \frac{1}{r_1 r_2}$$

$$\frac{1}{r_2} = r_1\, e^{2L\gamma}$$

$$r_2 = \frac{1}{r_1\, e^{2L\gamma}} = \frac{1}{1 \times e^{2 \times 15 \times 0.0005}} = \frac{1}{e^{0.015}} = \frac{1}{1.015} = 0.985 = 98.5\%$$

The reflectivity of other cavity mirror is r_2 = 98.5 %.

Example 4 : What will be the reflectivity of first cavity mirror if the reflectance of second mirror is 97% ? The length of the cavity is 15 cm and gain factor of laser material is 0.0005 per cm. **(Oct. 2017)**

Solution : Given : r_2 = 97% = 0.97, L = 15 cm, γ = 0.0005 per cm, r_1 = ?

Formula : Gain factor of laser material is,

$$\gamma = \frac{1}{2L} \ln \frac{1}{r_1 r_2}$$

$$2L\,\gamma = \ln \frac{1}{r_1 r_2}$$

$$e^{2L\gamma} = \frac{1}{r_1 r_2}$$

$$\frac{1}{r_1} = r_2\, e^{2L\gamma}$$

$$r_1 = \frac{1}{r_2\, e^{2L\gamma}}$$

$$r_1 = \frac{1}{0.97 \times e^{2 \times 15 \times 0.0005}} = \frac{1}{0.97 \times e^{0.015}} = \frac{1}{0.97 \times 1.0151} = \frac{1}{0.9845}$$

$$\therefore \qquad r_1 = 1.015 = 101.5\% \approx 100\%$$

The reflectivity of first cavity mirror is r_1 = 100 %.

Exercise

(A) Multiple Choice Questions :

1. Principle of laser is ......
   - (a) spontaneous absorption
   - (b) simulated emission
   - (c) both b and a

2. The purpose of the optical resonator in a laser is ......
   - (a) to provide cover to the active medium
   - (b) to provide path for atoms
   - (c) to provide selectivity of photons
   - (d) to send laser in specified direction

3. The life time of an atom in a metastable state is of the order of ......
   - (a) few seconds
   - (b) unlimited
   - (c) a nanosecond
   - (d) few milliseconds

4. The life time of an atom at the ordinary excited state is of the order of ......
   - (a) few milliseconds
   - (b) few nanoseconds
   - (c) few microseconds
   - (d) unlimited

5. Emission of a photon, by an excited atom due to interaction of external energy, is called ......
   - (a) Spontaneous emission
   - (b) Stimulated emission
   - (c) Induced absorption
   - (d) Light amplification

   Answers : (1) c, (2) d, (3) d, (4) b, (5) d.

(B) Short Answer Questions :

1. Define gain coefficient.
2. Define round trip gain
3. Define threshold gain.
4. State the condition for steady state oscillation.
5. State the condition for critical population inversion.
6. What is an optical feedback?
7. Draw the block diagram for optical resonator.
8. State lasing condition under critical population inversion.

(C) Long Answer Questions :

1. Write a short note on optical feedback.
2. Obtain the threshold condition for laser action.
3. Explain how cavity resonance frequency is obtained.
4. Write a short note on optical resonator.

Chapter 4

Laser Output

Objectives ...

- To study lifetime broadening.
- To study collision broadening.
- To study Doppler broadening.

4.1 Laser Modes

No light source is truly monochromatic. Every source emits light with a range of frequencies. Even a laser emits light with some range of frequencies, although that range can be very small. The causes of spectral line broadening are varied. The output laser consists of a number of very closely spaced, discrete frequency components (very narrow spectral lines) covering a moderately broad spectral range. The discrete components are called laser modes and the spectral range they occupy is approximately the fluorescent linewidth of atomic transition giving rise to the laser output.

The output of laser beam actually consists of a number of closely spaced spectral lines of different frequencies in a broad frequency range. The discrete spectral components are termed as laser modes, and coverage range is the line width of the atomic transition responsible for the laser output. Laser modes are categorized into axial and transverse modes.

1. The axial modes are constructed by the light waves moving exactly parallel to the cavity axis. Light incident on a mirror and that reflected from the mirror construct a standing wave similar to a string bounded at both the ends. All the axial modes are due to the propagation of plane waves along the line joining centers of two reflecting mirrors.

2. Unlike the plane waves propagating along the axis of the cavity in axial modes, there are some other waves traveling out of the axis that are not able to repeat their own path termed as transverse electromagnetic waves.

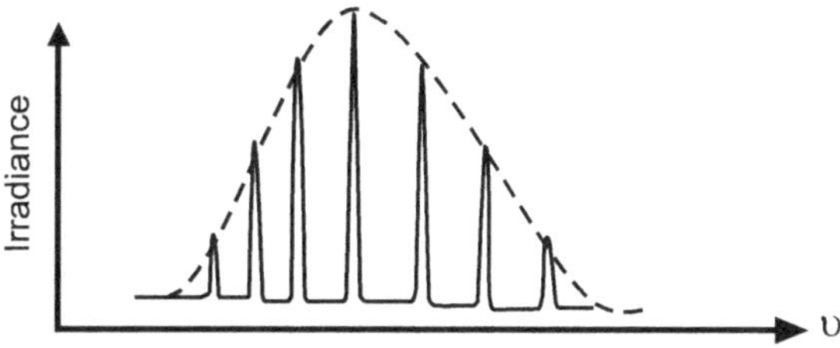

Fig. 4.1 : The axial mode of laser

4.2 Line Broadening (Nov. 2016, April 2017, Oct. 2017)

Broadening in laser physics is a physical phenomenon that affects the spectroscopic line shape of the laser emission profile. The laser emission is due to the excitation of an atom system between an excited state and a ground state. The difference in energy between these two states is proportional to the frequency or wavelength of the photon emitted and it is indeterminacy. The frequency/wavelength of the beam will have a certain width i.e. broadening.

Frequency at which absorption or emission of light takes place is given by

$$E = h\upsilon \qquad \qquad \text{... (4.1)}$$

where, ΔE is the difference between ground state energy and excited state energy of atom,

h is Plank's constant, and

υ is the frequency of the photon.

Depending on the nature of the indeterminacy, there can be two types of broadening.

1. Homogeneous broadening : A line broadening mechanism is referred to as homogeneous when it broadens the line of each individual atom, and therefore the whole system, in the same way e. g. natural broadening, collision broadening, etc. The most important homogeneous line broadening which is collision broadening in a gas, it is due to collision of an atom with other atoms, ions free electron, or the walls of the resonator. In a solid, it is due to the interaction of the atom with the phonon of the crystal lattice.

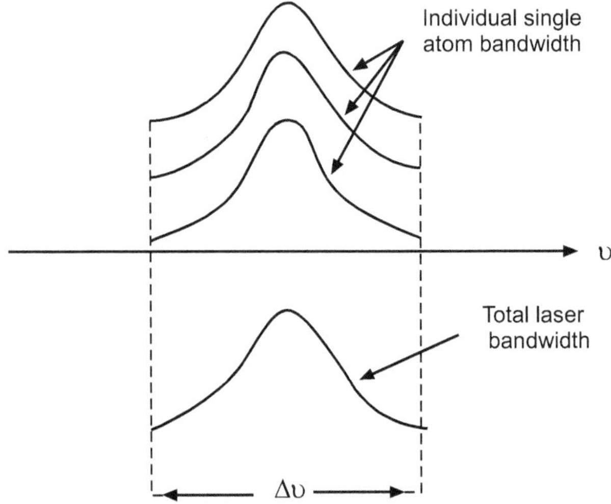

Fig. 4.2 : Homogeneous lines broadening

2. Inhomogeneous broadening : A line broadening mechanism is referred to as inhomogeneous when it leads to the atomic resonance frequencies being distributed over a band of frequencies and therefore results in a broadened line for the system as a whole without broadening the line of individual atoms e.g. Doppler broadening. Doppler broadening results from the difference in frequencies measured for the radiation emitted from atoms as they travel away from or towards an observer.

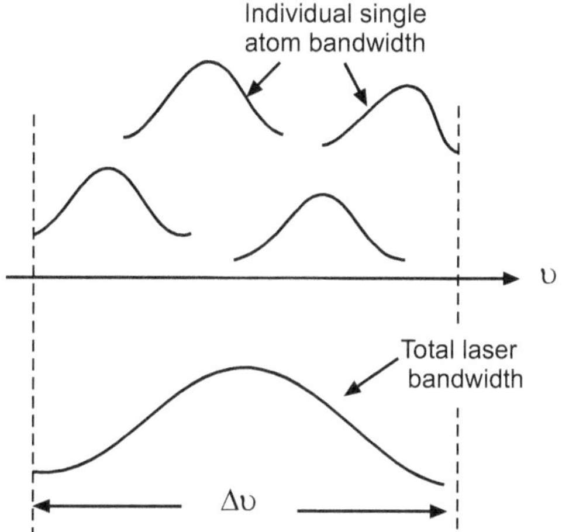

Fig. 4.3 : Inhomogeneous lines broadening

The total lineshape for a system can be easily measured through absorption spectroscopy. In fact, emission and absorption depend on photon frequency and are characterized by a linewidth. Line broadening has following three examples :

1. Natural (lifetime) broadening,

2. Collisional (pressure) broadening, and

3. Doppler broadening.

4.2.1 Natural (Lifetime) Broadening   (April 2016)

Natural broadening is an example of homogeneous broadening. Experimentally it is possible to reduce the effects of both Doppler broadening and Collision broadening by reducing the temperature of the gas, and by reducing its pressure, respectively. By taking data at different temperatures and pressures they can both be effectively eliminated. However, even in the limit of low temperature and pressure, the width of the spectral line is still not zero. There is a finite limit set not by the environment (i.e. temperature and pressure), but by the atoms themselves. This is referred to as the natural line width. The wavelength of an emitted photon depends on the energies of the upper and lower levels is,

$$h\upsilon = \frac{hc}{\lambda} = E_U - E_L \qquad \qquad \text{... (4.2)}$$

Note that energy levels in an atom are in reality not sharp. If energy levels are sharp, then energy difference is $E_U - E_L = 0$.

According to uncertainty principle,

$$\Delta E \cdot \Delta t \leq \hbar \qquad \qquad \text{... (4.3)}$$

where, $\hbar$ is $\dfrac{h}{2\pi}$.

This indicates that $\qquad \Delta E \propto \dfrac{1}{\Delta t} \qquad \qquad \text{... (4.4)}$

Therefore, we conclude that energy difference is 0 if $\Delta t = \infty$.

Hence, energy level has spread in terms of frequency because reciprocal of t is nothing but frequency.

$$\therefore \qquad \qquad \Delta E \propto \Delta \upsilon \qquad \qquad \text{... (4.5)}$$

where $\Delta \upsilon = \upsilon_1, \upsilon_2, \ldots\ldots, \upsilon_n$.

If life time of atom in energy state is $\Delta t = \dfrac{\hbar}{\Delta E} = 10^{-8}$ sec, then

$$\Delta \upsilon = \dfrac{1}{2\pi \, \Delta t} = 16 \text{ MHz}$$

This indicates that natural broadening is small in magnitude.

4.2.2 Collision (Pressure) Broadening   (April 2016, Oct. 2017)

Collision broadening is an example of homogeneous broadening. If atom which is emitting a wave train undergoes collision, then the phase of wave train is suddenly changed and effect produced shortening in the wave train is known as collision broadening. It is also called spectral line broadening.

Collision broadening results from atomic or molecular collisions in a gas or from interaction with lattice phonon in a solid. Collision broadening is homogeneous. Collisions cause random phase jumps in the wave functions. Alternatively, we can also think that the incoming electric field goes through phase jumps.

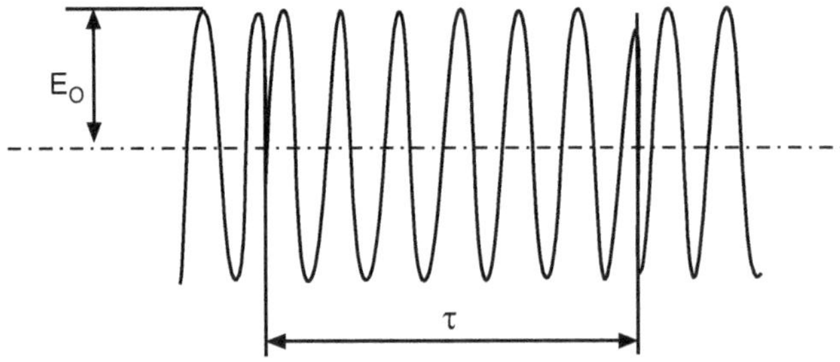

Fig. 4.4 : Collision broadening

If the absorbing atom responsible for an absorption line is suffering frequent collisions with other atoms, the electron energy levels will be distorted. This mechanism leading to broadening of emission and absorption lines is called collision broadening. The magnitude of this effect depends on the frequency υ_{col} at which such collision occurs. This frequency can be calculated as;

$$\upsilon_{col} = \upsilon_{th} \, \eta \, \sigma_{col} \qquad \ldots (4.6)$$

where, η is density of atoms,

σ_{col} is cross-sectional area of collisions, and

$\upsilon_{th} = \sqrt{2kT/m}$ is the thermal speed of atoms.

4.2.3 Doppler Broadening   (April 2017)

Doppler broadening is an example of inhomogeneous broadening. Doppler broadening is the broadening of spectral lines due to the Doppler Effect caused by a distribution of velocities of atoms. Different velocities of the emitting particles result in different Doppler shifts.

The frequency measured by observer increases if source & observer approach each other & decrease when they recent. In gas, atoms can move randomly, therefore observer can measure a range of frequencies. This broadening of line is called Doppler broadening. This broadening occurs under Doppler effects. The line shape function is

$$g\left(\upsilon_o\right) = \frac{1}{\Delta\upsilon} \qquad \ldots (4.7)$$

The thermal motion of absorbing atoms leads to small variations in the absorbed frequency υ due to the Doppler effect. If υ_o is the centroid frequency of the absorption line, then the Doppler effect will shift the frequency towards spectral line broadened.

$$\upsilon = \upsilon_o\left(1 + \frac{v_r}{c}\right) \qquad \ldots (4.8)$$

where v_r is the component of the velocity of the absorbing atom along the line of sight. Thus, the Doppler shift has an amplitude

$$\Delta\upsilon = \upsilon_o\frac{v_r}{c} \qquad \ldots (4.9)$$

4.3 Full Width at Half Maximum and Gain Bandwidth
(April 2016, Nov. 2016, Oct. 2017)

FWHM is the width of a spectrum curve measured between the two extreme values of the independent variable at which the dependent variable is equal to half of its maximum value. Half width at half maximum (HWHM) is half of the FWHM.

The limited range of frequency over which stimulated emission can provide sufficient gain is called **gain bandwidth** or gain. It is also called amplification bandwidth.

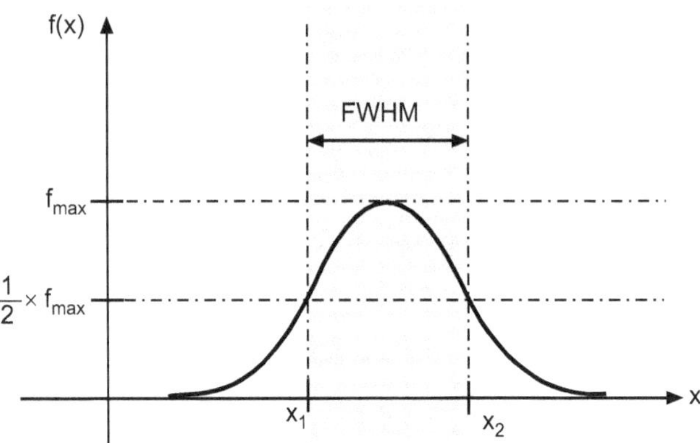

Fig. 4.5 : Full width at half maximum of spectrum

From equation (3.12), the spacing between neighbouring (m = 1) frequencies $\Delta\upsilon$ is,

$$\Delta\upsilon = \frac{c}{2\mu L} \qquad \qquad \text{... (4.10)}$$

By using the term $c = \upsilon\lambda$

$$\Delta\upsilon = \frac{\upsilon\lambda}{2\mu L} \qquad \qquad \text{... (4.11)}$$

$$\therefore \quad \frac{\Delta\upsilon}{\upsilon} = \frac{\lambda}{2\mu L}$$

$$\therefore \quad \frac{\Delta\lambda}{\lambda} = \frac{\lambda}{2\mu L}$$

$\therefore$ The half-width of gain profile of laser material device is

$$\Delta\lambda = \frac{\lambda^2}{2\mu L} \qquad \qquad \text{... (4.12)}$$

This indicates that laser will operate only at selected few frequencies for which the gain exceeds all the cavity losses.

Solved Examples

Example 1 : The half-width of gain profile of laser material device is 0.003 nm emitted wavelength of 6328 A°. Calculate maximum length of cavity in order to single mode of oscillation having refractive index is 1. **(Nov. 2016)**

Solution : Given : $\lambda = 6328$ A° $= 6328 \times 10^{-10}$ m, $\Delta\lambda = 0.003$ nm $= 0.003 \times 10^{-9}$ m ,

$$\mu = 1, \ L = ?$$

Formula : The half-width of gain profile of laser material device is

$$\Delta\lambda = \frac{\lambda^2}{2\mu L}$$

$\therefore \qquad L = \dfrac{\lambda^2}{2\mu\,\Delta\lambda}$

$\therefore \qquad L = \dfrac{(6328 \times 10^{-10})^2}{2 \times 1 \times 0.003 \times 10^{-9}} = 6673.83 \times 10^{-5} = 0.06674$ m

$$L = 6.674 \text{ cm}$$

Hence, maximum length of cavity is 6. 674 cm.

Example 2 : The half-width of gain profile of He-Ne laser material device is 0.002 nm having length of cavity is 10 cm. Calculate emitted wavelength of laser in order to single mode of oscillation having refractive index of material is 1.

Solution : Given : $\Delta\lambda$ = 0.002 nm = 0.002×10^{-9} m , μ = 1, L = 10 cm = 0.1 m, λ = ?

Formula : The half-width of gain profile of laser material device is

$$\Delta\lambda = \frac{\lambda^2}{2\mu L}$$

$\therefore \qquad \lambda^2 = 2\mu\,\Delta\lambda\,L$

$\therefore \qquad \lambda^2 = 2 \times 1 \times 0.002 \times 10^{-9} \times 0.1$

$\therefore \qquad \lambda^2 = 0.4 \times 10^{-12}$

$\therefore \qquad \lambda = 0.6324 \times 10^{-6}$ m = 6324×10^{-10} m or 6324 A°

Hence, emitted wavelength of laser is 6324 A°.

Exercise

(A) Multiple Choice Questions :

1. If coherence time (T_o) of wave train increases, then line width of laser will ......

 (a) decrease   (b) increase

 (c) be independent of T_o (d) depend on laser medium

2. The process when an atom absorbs a photon of light just equivalent to energy required to raise the atom to higher energy level is called ......

 (a) activation potential   (b) resonance potential

 (c) potential energy   (d) threshold potential

3. Probability of spontaneous emission increases rapidly with increase of ......

   (a) energy difference between the two states

   (b) frequency

   (c) both a & b

   (d) a or b

4. The method of raising a particle from lower state to higher energy state is called ......

   (a) pumping   (b) pumping source

   (c) population inversion   (d) none of these

Answers : (1) a, (2) b, (3) a, (4) a

(B) Short Answer Questions :

1. Define gain bandwidth.

2. Define FWHM.

3. State any one example of homogeneous broadening.

4. Is Doppler broadening an inhomogeneous broadening ?

5. Why natural broadening is also called as lifetime broadening ?

(C) Long Answer Questions :

1. What is lineshape broadening ? Explain homogeneous and inhomogeneous broadening.

2. Write a short note on Doppler broadening.

3. Write a short note on collision broadening.

4. What is line broadening ? Explain natural and collision broadening.

Chapter **5**...

Characteristics of Lasers

Objectives ...

- To understand and study various characteristics of laser beam.

- To know the difference between light & laser beam.

- To identify the most striking feature of laser beam.

5.1 Characteristics of Laser Radiation   (April 2015, 2017; Nov. 2016)

A laser is a device that emits light through a process of optical amplification based on the stimulated emission of electromagnetic radiation. The term "laser" is originated as an acronym for "Light Amplification by Stimulated Emission of Radiation". Light produced from the lasers has several valuable characteristics which are not shown by light obtained from other conventional light sources. These characteristics make them suitable for a variety of scientific and technological applications. The characteristics like monochromaticity, directionality, laser line width, brightness, and coherence of laser light make them highly important for various materials processing and characterization applications. These characteristics are discussed separately in the following subsections.

5.2 Directionality   (April 2016, 2017, 2018)

One of the most striking characteristics of laser is its directionality, that is, its output is in the form of an almost parallel beam. It can carry energy and data to very long distances for remote diagnosis and communication purposes. Conventional light sources emit radiation isotropically. Therefore, very small amount of energy can be collected using lens. The conventional light sources emit light in all directions. Therefore we need a narrow beam. Beam of an ideal laser is perfectly parallel and its diameter at the exit window is same as that after traveling very long distances, although in reality, it is impossible to achieve. Deviation in the parallelism of practical laser beam from the ideal is not due to any fault in the laser design, but due to diffraction from the edges of mirrors and windows. The directionality of a laser beam is expressed in terms of beam divergence. Light from a laser diverges very little. Upto a certain distance, the beam shows little spreading and remain essentially a bundle of parallel light rays. The distance from the laser over which the light rays remain parallel is known as Rayleigh range.

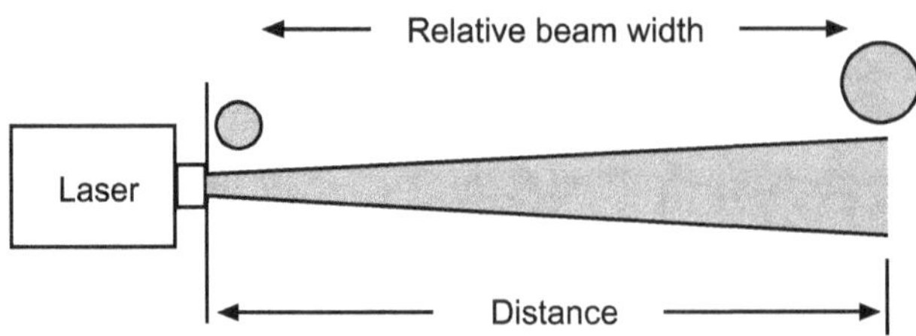

Fig. 5.1 : Laser beam divergence

There are two parameters which cause beam divergence. They are (i) size of the beam waist and (ii) Diffraction. Twice the angle of divergence is a full angle beam divergence that gives how much the beam will spread as it travels through space. The full angle divergence is given by,

$$2\theta = \frac{4\lambda}{\pi d_o} \qquad \qquad \text{... (5.1)}$$

where, $d_o = 2w_o$ is the diameter of the beam waist.

It is seen from equation (5.1) that the divergence is inversely proportional to d_o. Thus, divergence is large for a beam of small waist. The beam divergence due to diffraction is determined from Rayleigh's criterion,

$$\theta = 1.22 \frac{\lambda}{D} \qquad \qquad \text{... (5.2)}$$

where, D is the diameter of the laser's aperture.

A typical value of divergence for a He-Ne laser is 10^{-3} rad.

5.3 Monochromaticity   (April 2016, 2017, 2018)

The light emitted from a laser is monochromatic, that is, it is of one wavelength (colour). In contrast, ordinary white light is a combination of many different wavelengths (colours). If light coming from a source has only one frequency of oscillation, the light is said to be monochromatic and the source a monochromatic source. In practice, it is not possible to produce light having only one frequency. Light coming out of any source consists of a band of frequencies closely spaced around a central frequency υ_o. The band of frequencies $\Delta\upsilon$, is called the linewidth or bandwidth.

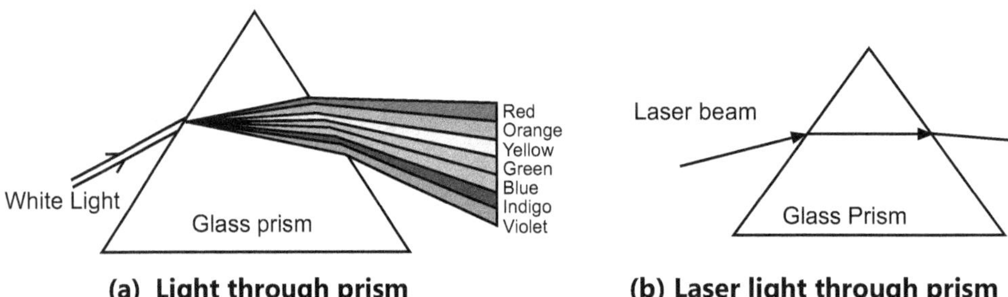

(a) Light through prism   **(b) Laser light through prism**

Fig. 5.2

The light from conventional sources has large linewidth of the order of 10^{10} Hz, but light from the lasers is more monochromatic having linewidth of the order of 100 Hz.

From the theory of interference, the linewidth of mirrored cavity is given by,

$$\Delta \upsilon = \frac{c}{2\pi L}\left(\frac{1-R}{\sqrt{R}}\right) \qquad \ldots (5.3)$$

where, L is the length of the cavity and R is the reflectance of the output mirror.

5.4 Coherence   (April 2018, 2017, 2016; Oct. 2017)

The light from a laser is said to be coherent, which means the wavelengths of the laser light are in phase in space and time. Lasers emit light that is highly directional. Laser light is emitted as a relatively narrow beam in a specific direction. Ordinary light, such as light coming from the sun, a light bulb, or a candle, is emitted in many directions away from the source. A laser differs from other sources of light. It emits light coherently, i.e. it maintains crest-to-crest and trough-to-trough correspondence.

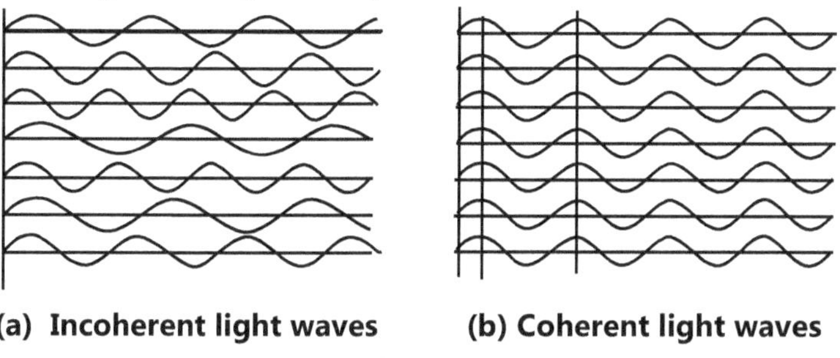

(a) Incoherent light waves   (b) Coherent light waves

Fig. 5.3

Coherence is a concept that establishes the limits within which a real light source can be considered ideal. Coherence requires that there is a connection between the amplitude and phase of the light at one point and time, and the amplitude and phase of the light at another point and time.

There are two types of coherence :   **(April 2016, Nov. 2016)**

1. Temporal (Longitudinal) coherence: (related to the emitted linewidth), and

2. Spatial (transverse) coherence: (related to the physical size of the source)

1. Temporal Coherence :

Temporal coherence is a characteristic of a single beam of light. Temporal coherence is a measure of the correlation between the phases of a light wave at different points along the direction of propagation. Temporal coherence tells us how monochromatic a source is. An excited atom in the process of passing to a lower energy state, gives up the excess energy. The transition lasts for a short time of 10^{-8} sec. In view of short distance in space and for a brief time, temporal coherence is characterized by two parameters namely coherence length l_{coh} and coherence time t_{coh}. Both the coherence length and coherence time measure how long light waves remain in phase as they travel in space. The coherence length depends on the central wavelength λ and the bandwidth, $\Delta\lambda$ of the wave packet. Assume our source emits waves with wavelength $\lambda \pm \Delta\lambda$. Waves with wavelength λ and $\lambda + \Delta\lambda$, which at some point in space constructively interfere, will no longer constructively interfere after some optical path length l_c, called the coherence length.

It is given by

$$l_{coh} = \frac{\lambda^2}{2\,\Delta\lambda} \qquad \ldots (5.4)$$

It can be expressed in terms of frequency as

$$l_{coh} = \frac{c}{\Delta\upsilon} \qquad \ldots (5.5)$$

and the coherence time t_c is

$$t_c = \frac{l_c}{c} \quad (\text{since } \lambda f = c) \qquad \ldots (5.6)$$

The coherence length is a very useful measure of temporal coherence because it tells us how far apart two points along the light beam can be and remain coherent with each other.

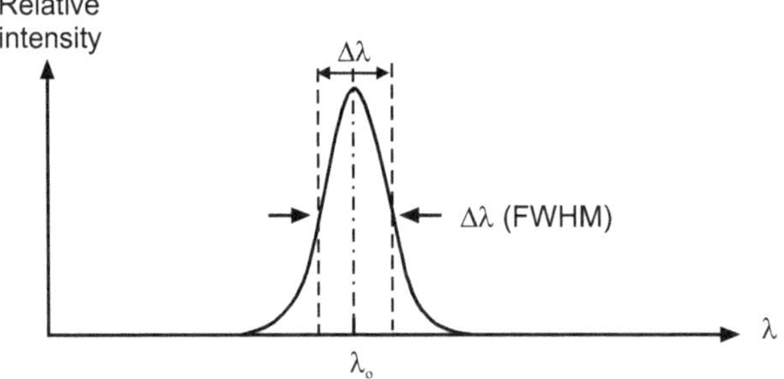

Fig. 5.4 : Temporal coherence

2. Spatial Coherence :

If the phase difference between the two electric vectors remains zero for all times, then the waves are said to be in spatial coherence. It implies that spatial coherence measures the area over which light is coherent. Spatial coherence is a measure of the correlation between the phases of a light wave at different point transverse to the direction of propagation. Spatial coherence tells us how uniform the phase of the wavefront is.

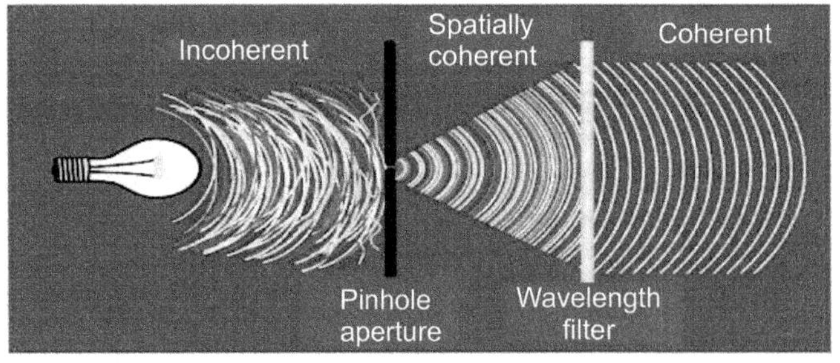

Fig. 5.5 : Spatial coherence

Coherence width is the distance along the wavefront (perpendicular to the direction of propagation), within which the amplitude and phase of the wave can be considered well defined and therefore predictable. Spatial coherence is determined by the coherence width (W_c).

$$W_c = k\left(\frac{\lambda}{D}\right) R \qquad \ldots (5.7)$$

where, k is a constant dependent on the shape of the source (for a circular source k = 1.22),

    D is the approximate diameter of the source, and

    R is the distance from the source.

An ideal point source would have infinite coherence width. Hence, the sharper the shadow gives the better spatial coherence of the source.

5.5 Intensity　　　　　(April 2016, 2017, 2018)

Intensity is the power of the laser beam divided by the cross-sectional area (Ω) of the beam. It is thus typically given in watts per square centimeter (W/cm^2). It is a measure of the amount of energy that can be applied to a specific region within a given amount of time. It is one of the two most important parameters in using the laser for materials processing applications such as welding, cutting, heat treating, ablating, and drilling, or for laser surgery. The other important parameter is the laser wavelength, since the amount of absorption of all materials, including biological materials, is dependent upon the wavelength of the light. In some instances a deep penetration of the beam is desired, for example in doing processes that must be carried out quickly. In that situation, a laser wavelength in which the material has a relatively low absorption would be selected. The power output of a laser may vary from a few milliwatts to a few kilowatts. But this energy is concentrated in a beam of very small cross-section.

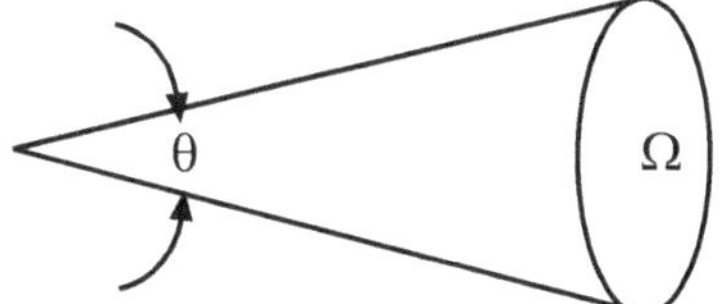

Fig. 5.6 : Laser beam brightness

The intensity of a laser beam is approximately given by,

$$I = \frac{10^2}{\lambda^2} P \qquad \ldots (5.8)$$

where P is the power radiated by the laser.

Radiance is a parameter that includes the beam intensity and takes into account the beam divergence angle. Laser beam divergence is usually given in milliradians because of the very low divergence of most lasers. Radiance becomes useful when a beam must be propagated over a reasonable distance before it is used or where the divergence can affect the focusing ability of the beam.

Solved Examples

Example 1 : Fluorescent tubelights emit visible light of wavelengths in the range of 4000 A° to 7000 A° with an average wavelength of 5500 A°. Calculate coherence length and coherence time. **(April 2017)**

Solution :　Given : λ = 5500 A°, Wavelengths in the range $\Delta\lambda$ = 4000 to 7000 A°,

l_{coh} = ? , t_c = ?

Formula : The coherence length is given by

$$l_{coh} = \frac{\lambda^2}{2\,\Delta\lambda}$$

$$= \frac{(5500)^2}{2 \times (7000 - 4000)}$$

$$l_{coh} = 5041.67 \approx 5042 \text{ A}°$$

The coherence time is

$$t_c = \frac{l_c}{c}$$

$$= \frac{5042 \times 10^{-10}}{3 \times 10^8}$$

$$= 1681 \times 10^{-18} \text{ sec}$$

Example 2 : Light from a sodium lamp, which is the traditional monochromatic source, has a coherence length of about 0.3 mm and bandwidth of about 6 A°. Calculate wavelength of sodium lamp. **(April 2018)**

Solution : Given : Bandwidth $\Delta\lambda$ = 6 A° = 6×10^{-10} m, l_{coh} = 0.3 mm = 0.3×10^{-3}, λ = ?

Formula : The coherence length is

$$l_{coh} = \frac{\lambda^2}{2\,\Delta\lambda}$$

$\therefore \qquad 2\,\Delta\lambda\,l_{coh} = \lambda^2$

$\therefore \qquad \lambda^2 = 2 \times 6 \times 10^{-10} \times 0.3 \times 10^{-3}$

$$= 0.36 \times 10^{-12}$$

$\therefore \qquad \lambda = 0.6 \times 10^{-6} = 6000 \times 10^{-10}$ m

$\therefore \qquad \lambda = 6000$ A°

Example 3 : Light from a He-Ne laser, which is the traditional monochromatic source, has a coherence length of about 100 m and wavelength 6328 A°. Calculate bandwidth of He-Ne laser. **(Nov. 2016)**

Solution : Given : l_{coh} = 100 m = 100×10^{-10} A°, λ = 6328 A° = 6328×10^{-10} m,

bandwidth $\Delta\lambda$ = ?

Formula : The coherence length is

$$l_{coh} = \frac{\lambda^2}{2\,\Delta\lambda}$$

$\therefore$
$$\Delta\lambda = \frac{\lambda^2}{2 l_{coh}}$$

$\therefore$
$$\Delta\lambda = \frac{(6328 \times 10^{-10})^2}{2 \times (100 \times 10^{-10})}$$

$\therefore$
$$\Delta\lambda = 2 \times 10^{-5} \text{ A}°$$

Example 4 : Calculate intensity of a He-Ne laser beam having emissive power 1 mW and wavelength 6328×10^{-10} m and intensity of ruby laser beam having emissive power 5 mW and wavelength 6943×10^{-10} m.　　　　　**(April 2016, Oct. 2017)**

Solution : Formula : The intensity of a laser beam is,

$$I = \frac{10^2}{\lambda^2} P$$

(a) For He-Ne laser beam : λ = 6328×10^{-10} m and P = 1 mW = 1×10^{-3} W

$$I = \frac{100 \times 10^{-3}}{(6328 \times 10^{-10})^2}$$

$\therefore$
$$I = 2.5 \times 10^{11} \text{ W/m}^2$$

(b) For ruby laser beam : λ = 6943×10^{-10} m and P = 5 mW = 5×10^{-3} W

$$I = \frac{100 \times 5 \times 10^{-3}}{(6943 \times 10^{-10})^2}$$

$\therefore$
$$I = 10.37 \times 10^{11} \text{ W/m}^2$$

Exercise

(A) Multiple Choice Questions :

1.　Laser beam is highly coherent, so it can be used in ....

　　(a)　interference　　　　　　　　(b)　diffraction

　　(c)　polarization　　　　　　　　(d)　scattering

2.　Which of the following is not true for LASER ?

　　(a)　Extremely intense light　　　(b)　Perfectly monochromatic

　　(c)　Incoherent　　　　　　　　(d)　Divergent

3.　Laser source is highly ....

　　(a)　coherent　　　　　　　　　　　　(b)　non-coherent

　　(c)　None of these

4.　Ordinary light emits ....

　　(a)　coherent light　　　　　　　　　(b)　incoherent light

　　(c)　None of these

5.　Laser source is highly ....

　　(a)　monochromatic　　　　　　　　(b)　polychromatic

　　(c)　None of these

6.　Laser beam divergence is usually measured in ....

　　(a)　millidegree　　　　　　　　　　(b)　milliradians

　　(c)　None of these

Answers : (1) a,　(2) c,　(3) a,　(4) b,　(5) a,　(6) b

(B) Short Answer Questions :

1.　What are the characteristics/properties of laser ?

2.　What is beam divergence ?

3.　Define coherence length.

4.　What is beam radiance ?

5.　State two types of coherence.

(C) Long Answer Questions :

1.　Explain the brief characteristics/properties of a laser beam.

2.　What do you understand by coherence ? Explain Temporal coherence and Spatial coherence.

3.　Write a short note on laser beam directionality.

4.　What are the characteristics/properties of laser ? Explain laser beam intensity.

5.　Calculate the coherence length for CO_2 laser whose line width is 1×10^{-5} nm of IR emission wavelength of 10.6 μm.

❑❑❑

Chapter 6

Types of Lasers

Objectives ...

- To observe various types of lasers.

- To describe working of various types of lasers.

- To study the applications of various types of lasers.

6.1 Types of Lasers

The first laser action was demonstrated in a ruby crystal by Maiman, in 1960. Since then, a large number of materials in various media have been found to give laser action at wavelengths in the visible, ultraviolet and infrared regions. These include various gases, solids, liquids, glasses, plastics, semiconductors, and dyes. In addition to the ruby crystal, many other crystals doped with rare earth ions have been found to give extremely good laser output. The numerous types and designs of lasers are increasing and can be broadly classified according to their production techniques.

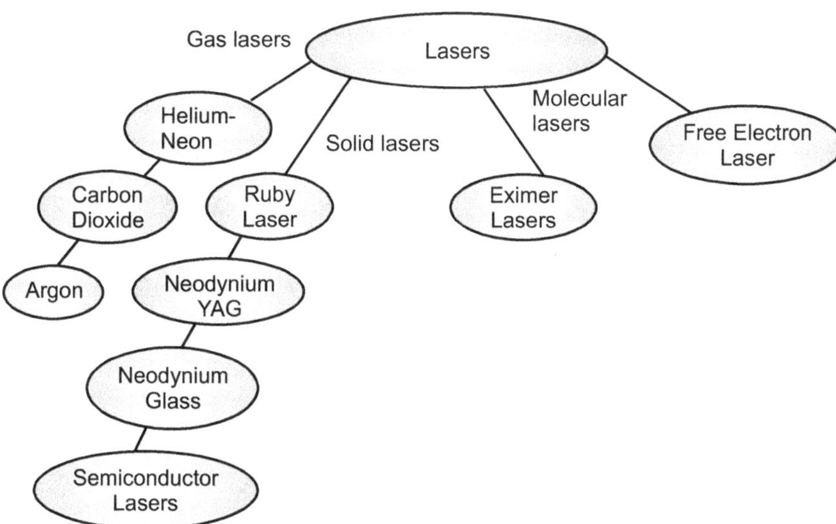

Fig. 6.1 : Broad categories of lasers

Some typical lasers and their emission wavelengths are as follows :

Laser Type	Wavelength (nm)
Argon fluoride (UV)	193
Krypton fluoride (UV)	248
Xenon chloride (UV)	308
Nitrogen (UV)	337
Argon (blue)	488
Argon (green)	514
Helium-neon (green)	543
Helium-neon (red)	633
Rhodamine 6G dye (tunable)	570-650
Ruby ($CrAlO_3$) (red)	694
Nd:Yag (NIR)	1064
Carbon dioxide (FIR)	10600

6.2 Solid State Laser　　　　　　(April 2017)

A solid state laser was invented in 1960. The solid state laser uses a solid crystalline material as the lasing medium. It uses a ruby rod (chromium doped aluminium oxide) with mirrors on both ends (one semitransparent) pumped with a helical xenon flash lamp surrounding the rod. It is optically pumped. The intense flash light is used to raise some of the chromium atoms to an upper energy state and participate in stimulated emissions. The result is an intense pulse of coherent red light at 694.3 nm wavelength.

Applications of solid state lasers :　　　　　　**(April 2017)**

Solid state lasers are used in materials processing (cutting, drilling, welding, marking, heat treating, etc.), semiconductor fabrication (wafer cutting, IC trimming), the graphic arts (high-end printing and copying), medical and surgical, rangefinders and other types of measurements, scientific research, entertainment, and many others.

6.2.1 Ruby Laser　　　　　　(April 2016, Nov. 2016)

Construction :

It is a solid state laser consisting of a pink ruby cylindrical rod whose ends are optically flat and parallel with a silvered and a partially silvered (50%) end. The rod is surrounded by a high intensity helical flash lamp filled with xenon gas which is intense enough to produce population inversion. Composition of Ruby is : Crystalline aluminium oxide (Al_2O_3 or host crystal) doped with 0.05% of chromium atoms (activator atoms). Al^{3+} ions are replaced by Cr^{3+} ions in crystal lattice. Cr^{3+} ions impart red colour to the white Al_2O_3 crystal.

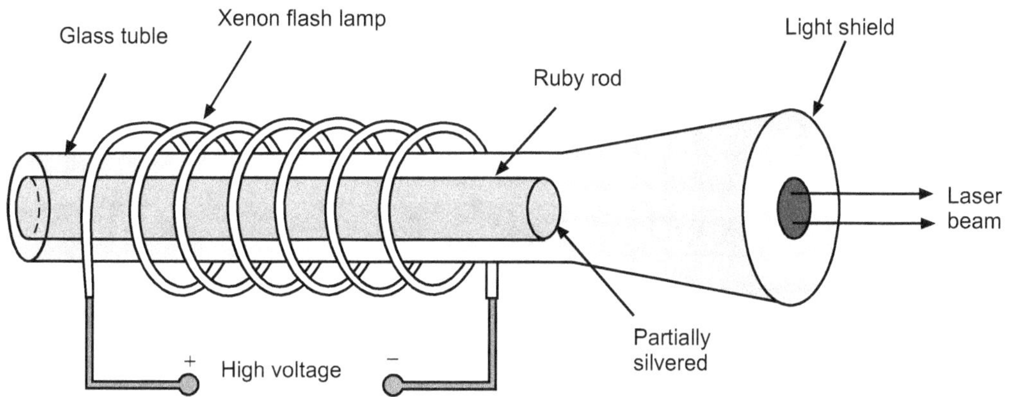

Fig. 6.2 : Typical diagram for ruby laser

Working :

The ruby laser is a three-level system. Chromium atoms consist of a metastable state of lifetime $\sim 3 \times 10^{-3}$ sec. When a flash of light of wavelength 550 nm falls upon the rod for a very short time (about a millisecond), the chromium ion, in the ground state, absorbs a photon and jumps to excited state E_3. The excited ions drop to the metastable state E_2 very soon as lifetime of ions in excited state is short. The transition is non-radiative, as the energy released is absorbed by the lattice in which it is absorbed and is dissipated as heat. But the number of atoms in metastable state goes on increasing as lifetime in metastable state is high and soon exceeds those in the ground state, thus bringing about population inversion.

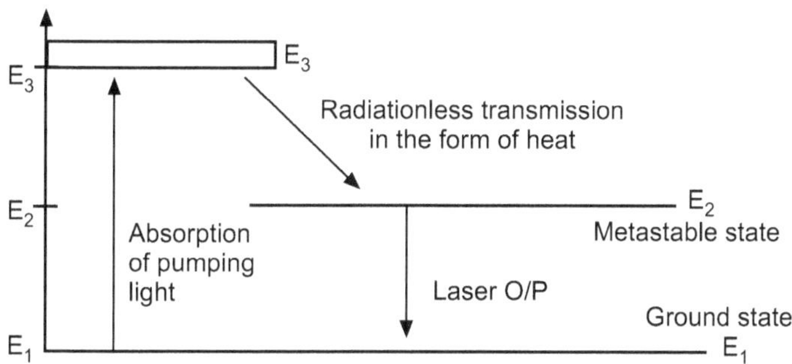

Fig. 6.3 : Energy level diagram for ruby laser system

After this state is achieved, one or two photons released due to spontaneous emission are sufficient to induce stimulated emission and light amplification will start. The transition from metastable state radiates photons, which after repeated reflection from the mirrors of the laser cavity amplifies largely to an intense beam.

An intense, highly directional, coherent beam of red light (λ = 694.3 nm) emerges from the partially silvered end of the ruby rod as laser beam.

Uses of Ruby Laser :

1. Due to low power, it can be used in toys for children.

2. It can be used in schools, colleges and universities for science programs.

3. It can be used as decoration piece and artistic display.

6.3 GAS LASERS　　　　　　　　　　　　　　　(Nov. 2016)

Gas lasers are widely available in the power of few milliwatts to few megawatts and wavelengths of UV-IR. They can be operated in pulsed and continuous modes. Based on the nature of active media, there are three types of gas lasers; atomic, ionic and molecular. Most of the gas lasers are pumped by electrical discharge. Electrons in the discharge tube are accelerated by electric field between the electrodes. These accelerated electrons collide with atoms, ions or molecules in the active media and induce transition to higher energy levels to achieve the condition of population inversion and stimulated emission.

6.3.1 Helium-Neon Laser　　　　　　　　(April 2017, Oct. 2017)

Construction :

It is a gas laser consisting of a mixture of helium (He) and neon (Ne) in a ratio of about 10 : 1 inside a narrow long discharge tube at pressure of 1 mm of mercury. The gas system is placed between a pair of plane mirrors or a pair of convex mirrors, out of which one is perfectly reflecting, while other is partially reflecting, forming the resonating system. The distance between the two mirrors is equal to an integral multiple of half wavelength of the laser light and supports standing wave pattern within the resonator system.

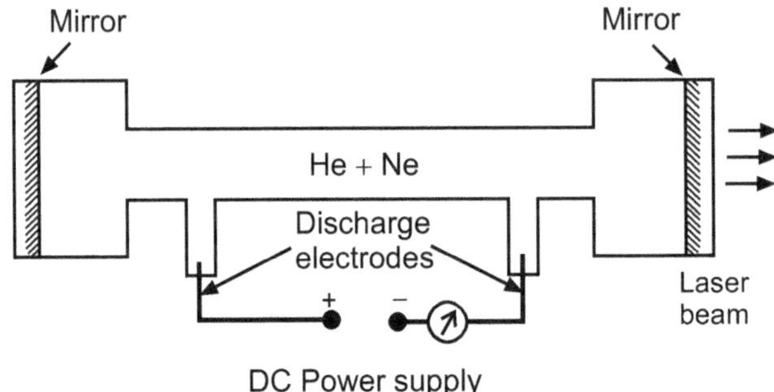

Fig. 6.4 : Typical diagram for He-Ne laser

Working :

Helium has three energy states. These are 3S, 2S and 1S, where 3S and 2S are metastable states. When an electrical discharge passes through the gas mixture, the helium atoms are excited by the impacts of accelerated electrons in the discharge tube due to its lower mass. As a result, some of the helium atoms are raised to its metastable states 2S and 3S from its ground state. The energy of the two excited states 2S and 3S of Ne is slightly less than the energy of the two metastable states of He atoms. Thus, after the collision of the excited helium atoms with neon atoms, the neon atoms in the ground state are raised to its 3S and 2S excited states and helium returns to its ground state by exchanging energy.

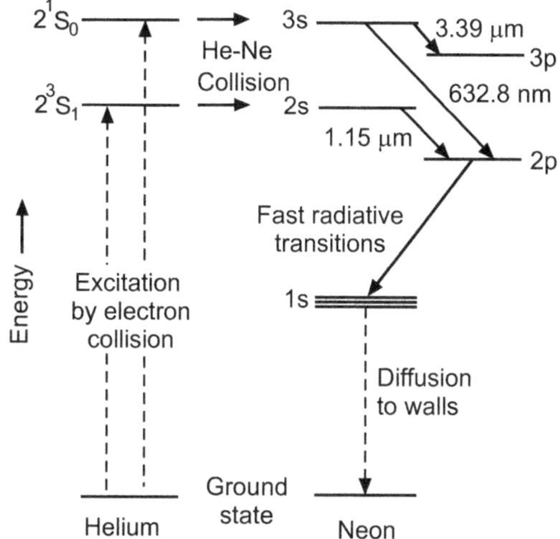

Fig. 6.5 : Energy level diagram for He–Ne laser system

The gas discharge process after some time leads population inversion in these metastable Ne (3S) and Ne (2S) levels relative to its lower 3P and 2P states. After achieving population inversion, one of the two photons released due to spontaneous emission can trigger stimulated emission and produce three types of lasing actions (3S · 3P, 3S · 2P, 2S · 2P). After that, the Ne atoms return to the lower laser levels 3P and 2P to the level 1S by spontaneous emission. From this level, Ne returns to ground state by collision with the walls of the tube. The cycle of events occur continuously as the discharge in the tube is maintained continuously. Thus it is known as continuous laser.

Uses of He-Ne Laser :

1. In interferometry,

2. In laser printing,

3. In barcode reading,

4. In holography,

5. For larger distance measurement, i.e. in laser modulation telemetry,

6. In the target aiming device used in guns.

6.3.2 CO_2 (Carbon Dioxide) Laser

CO_2 laser was one of the earliest gas lasers to be developed in BELL Labs in 1964 by Indian born C.K.N. Patel. The carbon dioxide gas laser is capable of continuous output powers above 10 kilowatts. It is also capable of extremely high power pulse operation. It exhibits laser action at several infrared frequencies, but none in the visible. Operating in a manner similar to the helium-neon laser, it employs an electric discharge for pumping, using a percentage of nitrogen gas as a pumping gas. The CO_2 laser is the most efficient laser, capable of operating at more than 30% efficiency. The carbon dioxide laser finds many applications in industry, particularly for welding and cutting.

Construction :

Carbon dioxide laser consists of a discharge tube having a diameter of 2.5 cm and a length of about 5 m. The discharge tube is filled with a mixture of carbon dioxide, nitrogen and helium gases in the ratio of 1 : 2 : 3 with water vapours. Pressures maintained are about P (for He) = 7 Torr, P (for N_2) = 1.2 Torr and P (for CO_2) = 0.33 Torr. The active medium is the CO_2, N_2 and He in the ratio of 1 : 2 : 3. The active centers are the carbon dioxide molecules because laser is achieved due to these molecules. The filling gas within the discharge tube consists of CO_2 (10–20%), N_2 (10–20%), H_2, He (the remainder of the gas mixture) and Xe (a few percent).

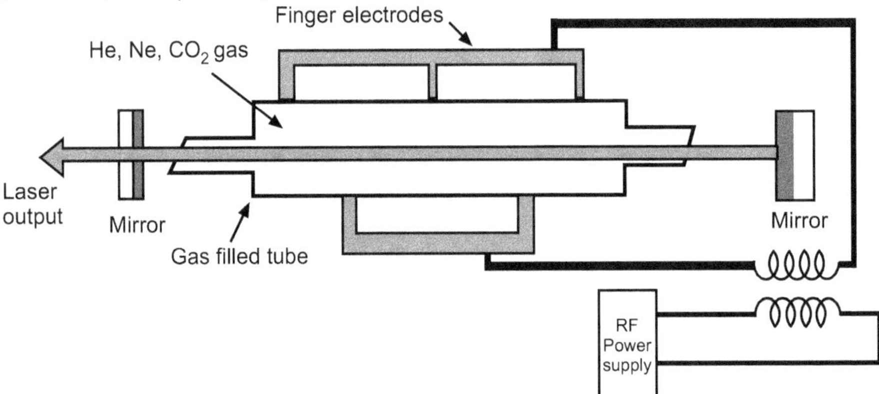

Fig. 6.6 : Typical diagram for CO_2 laser

Working :

Electric discharge method is used for pumping and achieving population inversion. In this method, electrons will collide with CO_2 molecules and pump them to excited states. The purpose of N_2 is to help in excitation of CO_2 molecules by colliding with CO_2 molecules and transferring the energy to them. So an N_2 molecule increases the pumping efficiency. All the gas mixtures are enclosed between a set of mirrors which forms the optical resonator system. One of the mirrors is completely reflecting and the other is partially reflecting.

Let us say the N_2 is excited from level F_1 to F_2.

N_2 goes to excited state by collision with electrons

$$N_2 + e_1 \rightarrow N_2^* + e_2$$

The excited N_2 transfers energy to CO_2 and CO_2 gets excited.

$$N_2 + CO_2 \rightarrow CO_2^* + N_2$$

The level F_2 happens to be metastable and thus the N_2 molecules excited to F_2 spend a sufficient amount of time before getting de-excited.　Operating temperature plays an important role in determining the output power of the laser.

Achievement of population inversion of CO_2 molecules :

The excited level (metastable level) of CO_2 molecules corresponds approximately to the same energy as the energy of the excited level F_2 of nitrogen. Thus when N_2 molecules in level F_2 collide with the CO_2 in the ground state E_1, an energy exchange takes place and this results in the excitation of CO_2 molecules to level E_5 and de-excitation of N_2 to the ground level F_1. Thus population inversion is achieved between vibrational levels E_5 and E_4 or E_5 and E_3. Thus E_5 is the upper laser level. E_3 and E_4 are lower laser levels.

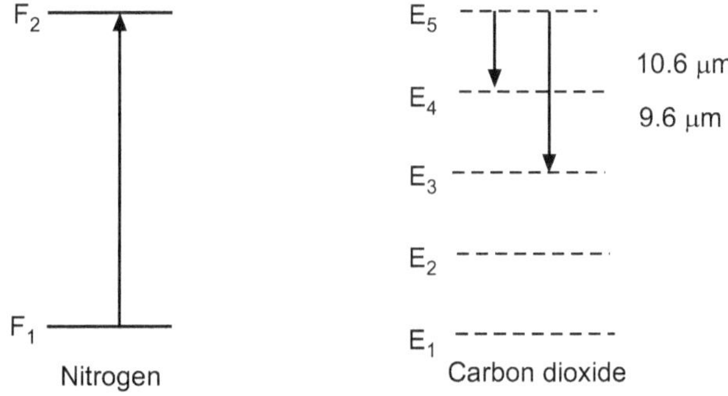

Fig. 6.7

Achievement of Laser :

Transitions produce lasers of wavelength 10.6 μm and 9.6 μm which lie in the far infra-red region. The CO_2 molecules in the states E_4 and E_3 de-excite to state E_2 through inelastic collision with unexcited CO_2 molecules.

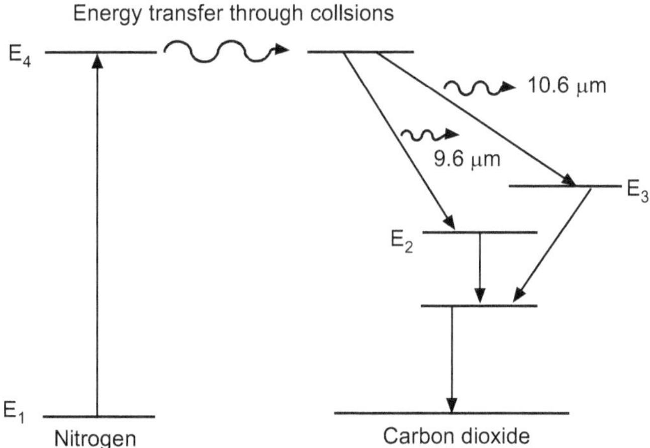

Fig. 6.8 : Energy level diagram for CO_2 laser system

This process is very fast. So there will be accumulation of CO_2 in this level and they can break the population inversion in upper levels because there is probability of excitation of molecules from E_2 to E_3 and/ or E_4. To stop the accumulation of CO_2 molecules in E_2 special additives like He and water vapours are added into the gas mixture. CO_2 molecules return to the ground state E_1 through collisions with the He to which it transfers the excitation energy. Other function of He is to conduct the heat away to the walls keeping CO_2 cold, this is because helium has high thermal conductivity.

Applications :

1. It is widely used in material processing such as drilling, cutting, welding, melting and annealing.
2. It is widely used in open air communication.
3. It is used in pollution monitoring and remote sensing.
4. It is used in medical field to perform microsurgery and bloodless surgery.

Merits and Demerits :

1. Outuput power can be increased by increasing the length of the gas tube.
2. Output power of the laser depends upon the operating temperature.
3. Dissociation of CO_2 molecule into CO may contaminate the active medium.

6.4 Advantages of Gas Laser over Solid State Laser

1. The light from gas laser has high degree of monochromaticity and directionality than that from solid state laser. This is due to imperfection in the crystal, thermal distortion and scattering.
2. The solid state laser needs cooling in time of operation, while the gas lasers can operate continuously without any cooling.

6.5 Liquid Lasers　　　　　　　　　　　　　　　　　(April 2016)

Efficiency, tunability and high coherence are the properties of dye lasers. Liquid lasers are optically pumped lasers in which the gain medium is a liquid at room temperature. The most successful of all liquid lasers are dye lasers. These lasers generate broadband laser light from the excited energy states of organic dyes dissolved in liquid solvents. Output can be either pulsed or CW and spans the spectrum from the near-UV to the near-IR, depending on the dye used. The large organic molecules of the dye are excited to higher energy states by arc lamps or flash lamps. The dye solution is pumped through the laser cavity. They are used in scientific, medical and spectroscopic research.

6.5.1 Tunable Dye Laser　　　　　　　　　　　　　　(April 2017)

The dye lasers are less expensive than the solid-state lasers and are relatively easy to maintain for regular operation. A tunable dye laser uses an organic dye (liquid solution) as the lasing medium. It consists of a host material such as alcohol in which the dye molecules (such as rhodamines or coumarins) are dissolved at a concentration of one part in ten thousand. Dyes exhibit a very high degree of fluorescence, i.e., when the dye is exposed to ultraviolet light, it glows with characteristic colour depending on the nature of the material. Different dyes have different emission spectra or colours. As a result, dye lasers cover a broad wavelength range from the ultraviolet at 320 nm to the infrared at about 1500 nm. A unique property of dye lasers is the broad emission spectrum (typically 30-60 nm) over which the gain occurs. When this broad gain spectrum is combined with a diffraction grating or a prism as the cavity mirrors, the dye laser output can be a very narrow frequency beam (10 GHz or smaller). Frequency tuning over even larger ranges is accomplished by inserting different dyes into the laser cavity.

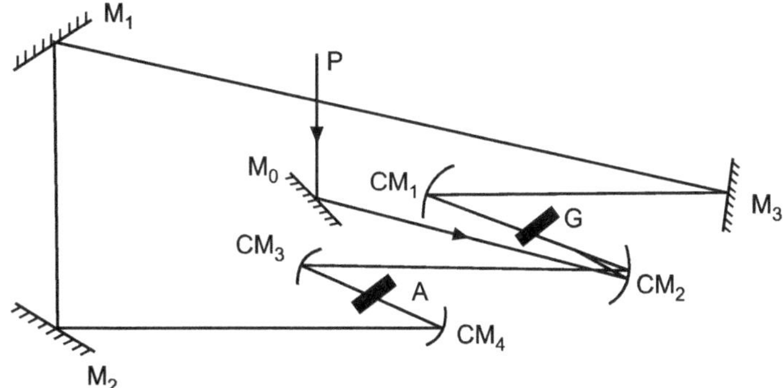

Planar mirrors : M$_0$, M$_1$, M$_2$, M$_3$,   Curved mirrors : CM$_1$, CM$_2$, CM$_3$, CM$_4$

P - Pump laser beam, A - Saturable absorber dye jet

G - Gain dye jet, OC - Output coupler

Fig. 6.9 : Typical diagram for ring type tunable dye laser

A ring laser design is often chosen for continuous operation. The mirrors of the laser are positioned to allow the beam to travel in a circular path. The dye molecule is usually very small. Sometimes a dye jet is used to help in avoiding reflection losses. The dye is usually pumped with Nd : YAG laser. The liquid is circulated at very high speeds, to prevent triplet absorption from cutting off the beam. A ring laser does not generate standing waves which cause spatial hole burning. This leads to a better gain from the lasing medium.

Applications :

1. Dye lasers are used for studying the properties of a material when the laser frequency is varied over a wide range.

2. Dye lasers become important for spectroscopy, photochemistry, pollution monitoring, isotope separation, etc.

3. Dye lasers are used for producing ultra short optical pulses by a technique known as mode locking.

4. Dye lasers are also used in optical communication.

6.6 Semiconductor Laser

Semiconductor lasers are based on semiconductor gain media. Optical gain is usually achieved by stimulated emission under conditions of a high carrier density in the conduction band. The semiconducting lasers are also called junction lasers or junction diode lasers because they produce laser energy at the junction of two types of impurities in a semiconductor. They are also called injection lasers because electrons are injected into the junction region. The laser consists of a semiconducting crystal such as gallium arsenide, lead selenide, etc, with parallel faces at the ends to serve as partially reflective mirrors. Gallium and arsenic compounds are used to generate infrared rays when the current is passed through them. This implies that these semiconductors convert electrical energy into photons. But, these were ordinary incoherent light rays and were not produced by the laser action. However, when the gallium-arsenide crystal is injected through it, then the laser action does take place.

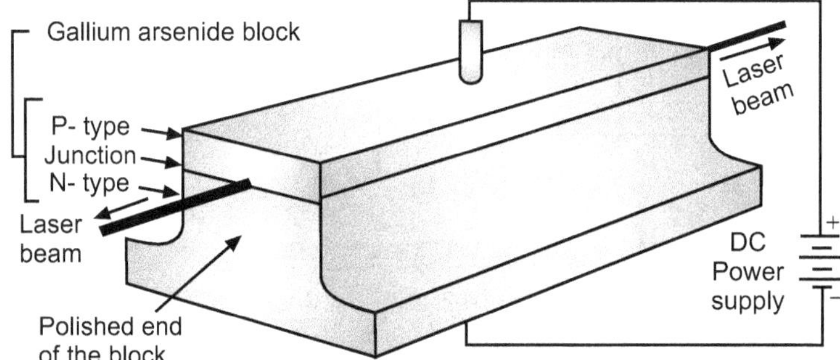

Fig. 6.10 : Typical diagram for semiconductor laser

Applications :

1. Semiconductor lasers are very suitable for applications where high powers are not required.

2. They are used in the area of communication in which the near-infrared laser beams can be transmitted over long distances through low loss optical fibres.

3. They have a large market as reading devices for compact disc players.

Exercise

(A) Multiple Choice Questions :

1. Pumping source inside CO_2 laser is .... pumping.

   (a) optical　　　　　　　　　(b) electrical

   (c) chemical　　　　　　　　(d) X-ray

2. The reconstruction process in holography involves .... phenomenon.

   (a) Interference　　　　　　　(b) Diffraction

   (c) both (a) & (b)　　　　　　(d) none of these

3. Pumping source preferred for gaseous lasers is .... pumping.

 (a) optical   (b) electricaL

 (c) chemical   (d) X-Ray

4. Minimum value of energy in eV required to excite an e^- from ground state to higher state corresponds to .... potential.

 (a) Excitation   (b) ionization

 (c) critical   (d) atomic excitation

5. In Ruby laser the atoms are excited by ....

 (a) ruby rod   (b) flash tube

 (c) silvered mirror   (d) semi-transparent mirror

6. In a Ruby laser, a large number of atoms occupy .... state.

 (a) ground   (b) excited

 (c) metastable   (d) normal

7. Laser can be used in the ....

 (a) fission reaction   (b) thermo-nuclear fusion

 (c) production of white light   (d) none of these

8. What colour does the Ruby laser emit?

 (a) Pure white light   (b) Pure blue light

 (c) Pure red light   (d) Pure yellow light

9. What is the low powered carbon dioxide laser used for?

 (a) Removal of tattoos   (b) Removal of melanomas

 (c) Removal of port wine stains   (d) All of these

10. In Ruby laser, the active medium is ....

 (a) solid   (b) liquid

 (c) gas   (d) semi-liquid

11. In Ruby laser, the activator atom is ....

 (a) aluminium   (b) chromium

 (c) oxygen   (d) rubidium

12. In Ruby laser, the host crystal is ....

 (a) Al_2O_3 (b) MnO_2

 (c) $CaCO_3$ (d) None of these

13. In He-gas laser, the ratio of mixture of Ne gas and He gas is ………. .

   (a) 1 : 10　　　　　　　　　　　　(b) 10 : 1

   (c) 11 : 10　　　　　　　　　　　 (d) 10 : 11

14. In He-Ne laser, He has ………. .

   (a) 3 energy states　　　　　　　　(b) 2 energy states

   (c) 4 energy states　　　　　　　　(d) None of these

15. In He-Ne laser, two energy states of Ne are　………. .

   (a) 1S, 2S　　　　　　　　　　　　(b) 1S, 3S

   (c) 2S, 3S　　　　　　　　　　　　(d) 2S, 2S

Answers : (1) a, (2) a, (3) a, (4) a, (5) b, (6) c, (7) a, (8) c, (9) b, (10) a, (11) b, (12) a,

　　　　 (13) a, (14) a, (15) c

(B) Short Answer Questions :

1. Enlist various applications of lasers.
2. State various types of gas lasers.
3. Give two types of solid state lasers.
4. Give two applications of solid state lasers.
5. Enlist various applications of gas lasers.
6. Give two advantages of gas lasers over solid state laser.
7. Give two applications of liquid lasers.
8. State two types of liquid lasers.
9. Give two applications of semiconductor lasers.
10. Draw typical diagram for ring type tunable dye laser.
11. Draw energy level diagram for CO_2 laser.

(C) Long Answer Questions :

1. Describe construction and working of Ruby laser. Also give its applications.
2. Describe construction and working of He-Ne laser. Also give its applications.
3. Write a short note on tunable dye laser.
4. Write a short note on semiconductor laser.
5. Explain with neat diagram construction and working of CO_2 gas laser. Also give its applications.
6. Explain any one type of laser with neat diagram.

Chapter **7** ...

Applications of Lasers

Objectives:

- To know wide applications of lasers in industries.
- To know various applications of lasers in nuclear science fields.
- To study various applications of lasers in defence.
- To know various applications of lasers in medical science fields.
- To study various applications of lasers in optical science fields.

Introduction

Laser radiations were used in various applications immediately after the first Ruby laser was designed and constructed by Maiman in 1960. As early as 1961, its radiations were used to treat eye and skin diseases. Now it is used in medicine, astronomy, geodesy, metrology, chemistry, biology, spectroscopy, holography, power engineering, in various processes in engineering, as well as in communication technology, automation and remote control, in military technology, entertainment industry, art restorations, etc.

7.1 Lasers in Mechanical Industries   (Nov. 2016; April 2017, 2018)

Laser beam can be controlled and located much more precisely than an arc or a flame. The intense heat generated by its absorption in various materials is used for high precision welding, drilling, and cutting. No distortion is caused in the material welded by laser because it produces minimum shrinkage. Traditional welding has its limitations. Certain materials cannot be welded by the conventional means. These processes require transfer of energy from the laser beam to workpiece. It can happen only if the material has high absorption at the wavelength corresponding to the laser beam. Laser technology has enabled reliable welding of gold with silicon and germanium, aluminium with nickel, tantalum with copper and several other metals used in electronic equipments. With laser, it is even possible to join metallic and non-metallic materials.

(a) Laser Welding :

Laser welding means two or more pieces are joined together without impurities into the joint. The work-pieces do not get distorted as the total amount of power input is very small compared to conventional welding process. Unlike electron beam welding, it can be done in atmosphere. Laser welding is a contactless process. The laser beam heats the edges of the two plates to their melting points and causes them to fuse together. The heat affected zone is relatively small because of rapid cooling. Laser welding can be done even at the places where it is difficult to reach. The process is easily automated. CO_2 laser is the most common laser used in welding of sheets or films of plastic materials. Pulsed laser is also used for welding purpose. He, Ar and N_2 gases are often used in laser welding for protection against oxidation of metal surface. In automobiles, laser welding is used to weld the curved contours.

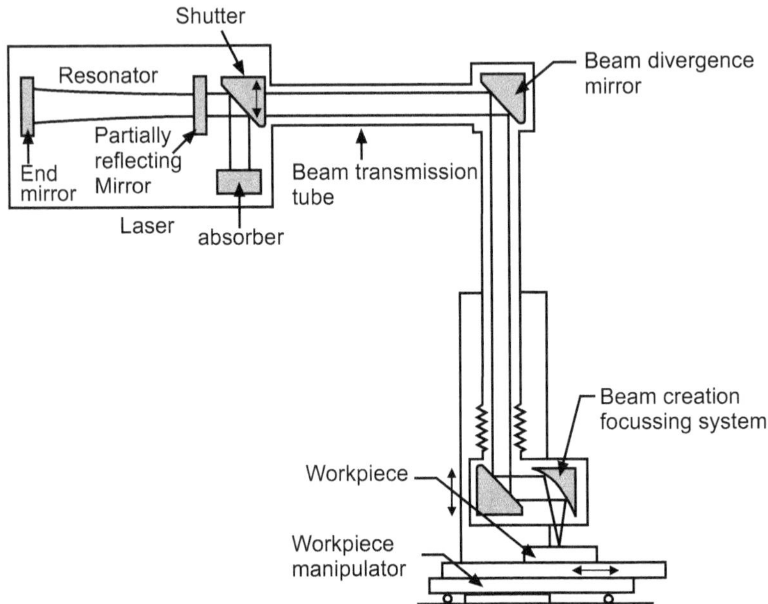

Fig. 7.1 : Schematic layout of CO_2 gas laser welding

(b) Laser Cutting :

Carbon dioxide gas laser is a continuous wave laser, because of that it is extensively used for cutting a wide range of materials such as graphite, diamond, tungsten, carbide, all metallic foils, ceramics, sapphire, paper, wood, cloth, glass, quartz, composites, steel and ferrite. In most cases, continuous cutting is carried out with assist gases like oxygen, carbon dioxide, or air, which produces both mechanical and chemical actions intensifying the thermal effects. This gas assisted cutting is applicable to the metals of thickness upto 5 mm with cut-widths down to 30 μm. The most promising field of laser cutting is the cutting of steels of small thickness such as several millimeters and also of non-metallic materials.

Use of laser cutters in the garment industry is very useful application of the lasers. With the aid of computers, lasers can cut clothing many times faster than the old techniques used by tailors. The laser system also consists of a computer, programmed with cutting instructions and patterns for various to garments. The cutting process essentially consists of removing material. Laser cutting is done with the assistance of air, oxygen or dry nitrogen gas jet. The role of the jet is to cool the surface of the material and blow away the debris from the cutting zone. The advantage of laser cutting is that it is fine and precise. It introduces a minimum mechanical distortion and thermal shock in the material being cut.

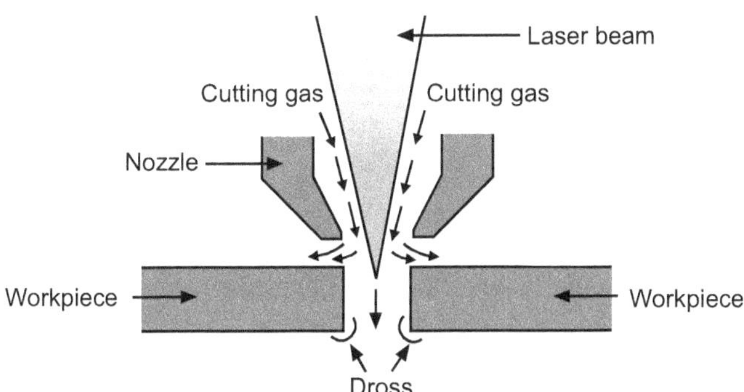

Fig. 7.2 : Schematic layout of CO_2 gas laser cutting

(c) Laser Drilling :

Laser drilling is a non-contact process and does not require a physical drill bit. Laser drilling of metals is based on a face heating phenomenon. The absorbed intensity is transformed into heat within the penetration depth of laser radiation. The illuminated spot on the surface reaches boiling temperature and material removal starts due to the processes of vaporization and melt expulsion. Laser enables drilling of a diamond die in a few minutes as against 20 hours taken by conventional methods. There is no wastage in the process and the saving in terms of the cost of diamond dust helps in recovering the financial outlay on such a drilling system. Laser light energy is primarily applied in effecting micro-openings in rubies and diamonds. Without heating the entire machined unit, now it is possible to drill filament canals in refractory materials. Laser drilling of holes in a diamond takes 2 to 3 minutes as against 2 to 3 days taken by conventional drilling. A laser installation in Russia drills holes with diameters and depths from 0.005 to 0.8 mm and depths from 1 to 3 mm, respectively, in diamonds of any size and shape. The plus point about laser drilling is that it does not cause any damage to the diamond or any other processed material.

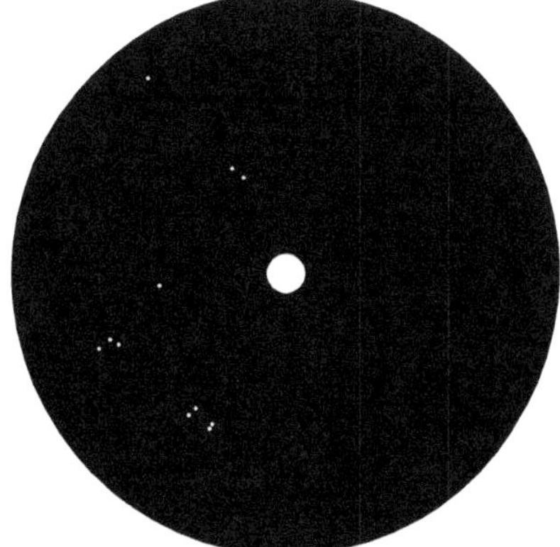

Fig. 7.3 : Typical hole drilled in a Ruby disc

For laser drilling, usually pulsed carbon dioxide, Nd:YAG and Alexandrite laser is used. Special operations like drilling of holes with diameter less than 0.5 mm using conventional techniques are difficult. However, the pulsed laser microdrilling is quite successful for such operations, both for metals and non-metals. The Nd:YAG laser emits at 1.06 μm. The superior performance of alexandrite laser is due to its shorter wavelength and continuous spiking of its output. CO_2 laser is equally suitable for drilling in metallic and non-metallic materials. Lasers are routinely used for drilling in ceramic materials.

7.2 Lasers in Nuclear Science

There are many applications of high power laser in nuclear astrophysics, accelerator physics, accelerator driven system, nuclear photonics, detection and characterization of nuclear material, nuclear medicine, isotope separation and nuclear fusion.

(a) Isotope Separation :

In the early 1970's, significant work began on the development of laser isotope separation technologies for uranium enrichment. The natural Uranium (U) has two principal isotopes such as U_{235} and U_{238}, which are used in fuel nuclear reactors. Isotopes are chemically identical. Tuning dye laser is used to separate it because it has very narrow linewidth. The U_{235} atoms can be ionized. The ionized U_{235} atom can be separated from the natural U_{238} atoms using electrostatic fields. Uranium-235 is of particular interest because it is the only fissile material that occurs in nature in significant quantity and it can be used to construct a nuclear explosive device if a sufficient quantity can be acquired. If the mixture of isotopes is irradiated by a source of narrow bandwidth; it is possible to excite one isotope without disturbing the other. U_{235} isotope sustains the fission reaction and it is useful for nuclear power generation. It is also used for production of atomic bombs. U_{238} isotope do not sustain the fission chain reaction.

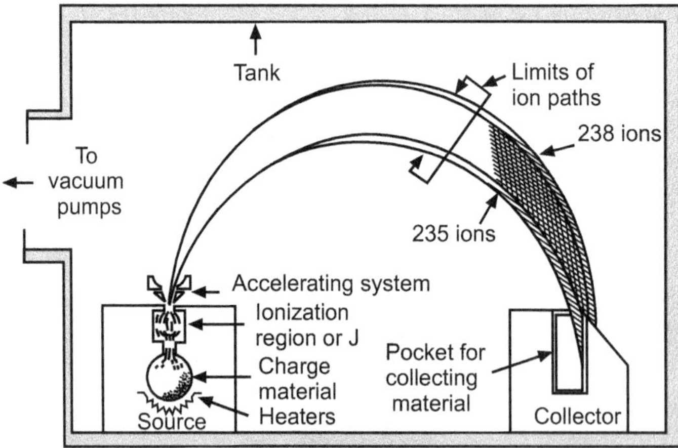

Fig. 7.4 : E-M method for isotope separation

The enrichment of natural uranium is very important process. This enrichment is performed using gaseous diffusion. It is very expensive and time consuming process. The differences in the nuclear mass shift slightly the electronic energy levels and each isotope absorbs light at different characteristic wavelengths. The absorption bands are fairly narrow and lie close to each other.

(b) Nuclear Fusion :

Nuclear fusion is process in which large amount of energy is released through the fusion of nucleus of lighter elements to form nuclei of heavier elements. It is the process by which nuclei of light elements such as deuterium (an isotope of hydrogen) are fused (or joined) together to produce heavier elements like helium. In this reaction, a large quantity of energy and neutrons are released. One million degrees centigrade temperature is required to take place thermonuclear fusion. Now-a-days, thermonuclear fusion is achieved by the implosion of the atoms of the material by a focused high energy laser beam.

Thermonuclear reaction has following advantages :

1.   The immense energy comes from a very small quantity of material.
2.   The supply of fusion fuel is virtually inexhaustible as deuterium can be extracted cheaply from the world's oceans.
3.   There is no problem of atmospheric pollution.
4.   It will be simpler and easier to make a hydrogen bomb which will be 'clean', i.e. its explosion will be free from the lingering effects of radioactive fallout.
5.   It offers a low cost.
6.   It produces less radioactive nuclear waste materials.
7.   The light nuclei required for a fusion reaction are available in abundances on earth than the heavy elements needed for fission.

Thermonuclear reaction has following disadvantages :

1.   Fusion reactions are very hard to produce and control.
2.   Extremely high temperature and pressures are required in order to make light nuclei to overcome their mutual repulsion and combine to release energy.

7.3 Lasers in Defence

There are many uses of lasers in military such as target designation and ranging, defensive countermeasures, communication and directed energy weapons.

Range Finder :   **(April 2016, 2017, 2018; Oct. 2017)**

Lasers are used for measuring distances, detecting distant objects and collecting information about them in radars. This is known as range finding. Neodymium and CO_2 lasers are range finding lasers. These laser range finders are light weight and have higher reliability and superior range accuracy as compared to the conventional range finders. In the optical range finding system, lasers transmit a signal in the form of pulses. This collimated pulse of the laser beam is directed towards a target and the reflected light from the target is received by an optical system and detected. The time taken by the laser beam for the to and fro travel from the transmitter to the target is measured. When half of the time thus recorded is multiplied by the velocity of light, the product gives the range i.e. the distance of the target.

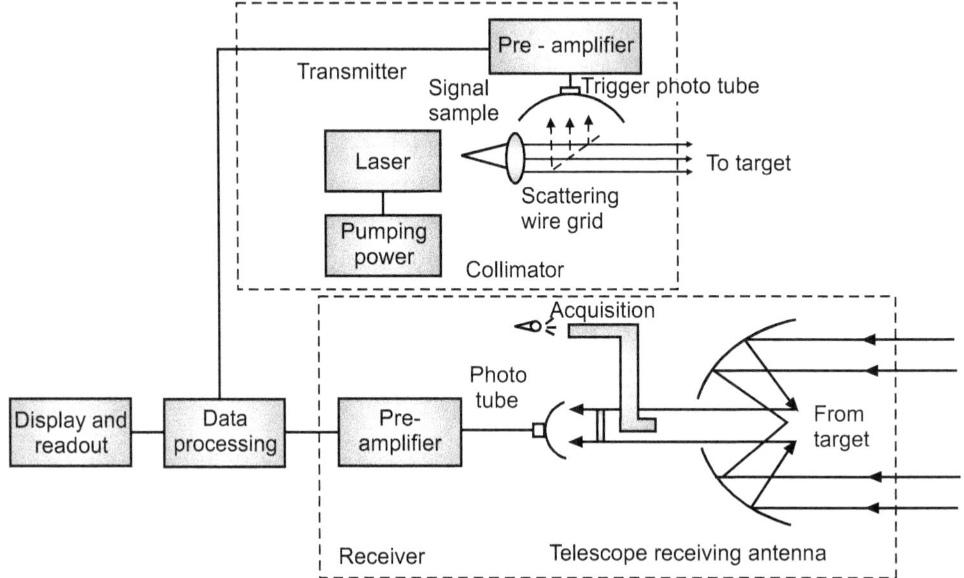

Fig. 7.5 : Typical block diagram for laser range finder

It is functionally divided into following four parts :

(i) Transmitter : The transmitter uses a Q-switched Nd:YAG laser which sends out single, collimated and short pulse of laser radiation to the target. A scattering wire grid directs a small sample of light from the transmitter pulse on to the photodetector, which after amplification is fed to the counter. This sample of light starts the counter.

(ii) Receiver : The reflected pulse, received by the telescope, is passed through an interference filter to eliminate any extraneous radiation. It is then focused on to another photodetector. The resulting signal is then fed to the counter. A digital system converts the time interval into distance.

(iii) Display and readout : The range, thus determined by the counter, is displayed in the readout.

(iv) Lighting telescope : The lighting telescope permits the operator to read the range while looking at the target.

Now-a-days, low flying aircrafts are used in ground attacks. They are fitted with laser device for measuring the range of the target, and guiding the bomb onto it thereby reducing the human element to the minimum. Laser powers of beam suffice for such type of operation. However, if it is intended to destroy the enemy weapons, high intensity laser beams having powers of megawatts are needed. The major problem involved in this is that as the laser beam travels through air, it is absorbed through to a small extent. The absorption of energy leads to heating up of the air. Therefore, random turbulence is created there around and the laser beam is bent away from the target. This makes it hard both to track the target and to focus the laser beam onto it effectively.

7.4 Lasers in Medical Science

Lasers are extensively used in medicine and they play important role both in diagnosis and surgery such as cosmetic surgery, removing tattoos, scars, stretch marks, sunspots, wrinkles, birthmarks, hairs, refractive surgery, soft tissue surgery, general surgery, gynecological, urology, laproscopic, laser therapy, removal of tumours, skin health assessments, tooth whitening, oral surgery, etc.

Eye Surgery :

Thousands of Canadians are turning to laser eye surgery to correct their vision and eliminate their dependency on glasses or contact lenses. Like all medical procedures, laser eye surgery provides benefits, but poses risks. Laser eye surgery is a medical procedure that involves the use of a laser to reshape the surface of the eye. This is done to improve or correct short-sightedness, long-sightedness and astigmatism. The first big success of lasers in medicine was in the treatment of eye. Argon laser was used to weld detached retina. Photocoagulated blood vessels are grown into the region in front of the retina which causes blocking of vision. Argon laser beam is focused on the desired point of the retina. The argon laser beam easily passes through transparent portions of the eye. Retina is a sensitive membrane inside the eyeball. Its detachment from the surrounding choroid coat initiates due to a hole in the retina caused by an injury or degenerative changes during the old age. This makes the thick fluid vitreous humour seep and fill itself between the retina and chord oat. The pressure between the retina and choroid coating damages the retina which soon gets detached from the optic nerve at the back to cause blindness.

7.5 Lasers in Optical Science

(a) Holography :   **(April 2016, 2017, 2018; Nov. 2016; Oct. 2017)**

Holography is the important application of laser beam. It is the production of true three-dimensional pictures in space without the use of lens. The record of this three-dimensional image of the object on a film is called a **hologram**. To get a high quality hologram, radiation of high coherence is required. He-Ne gas laser gives highly coherent light. Laser light is split into two beams. One beam called the reference beam is directed towards photographic plate. The other beam called object beam is directed towards the object to be recorded. The mixing of reference beam and the reflected beam produces interference pattern on photographic plate.

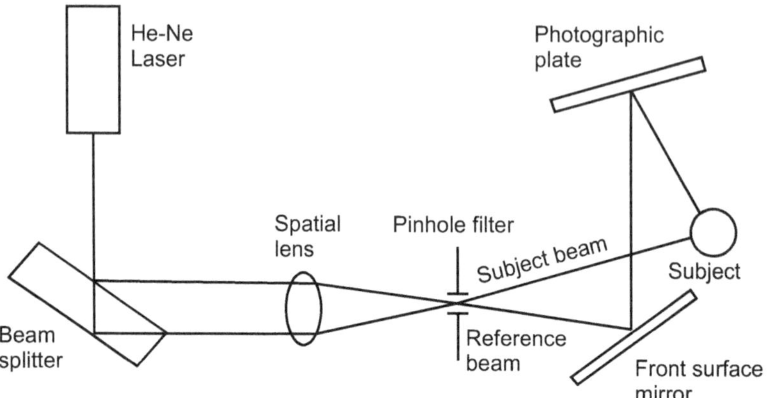

Fig. 7.6 : Typical block diagram of hologram

There are three types of holograms. These are

1. **Reflection holograms,** in which a truly three-dimensional image is seen near its surface.

2. **Transmission holograms,** viewed with laser light.

3. **Hybrid holograms,** between the reflection and transmission types of holograms.

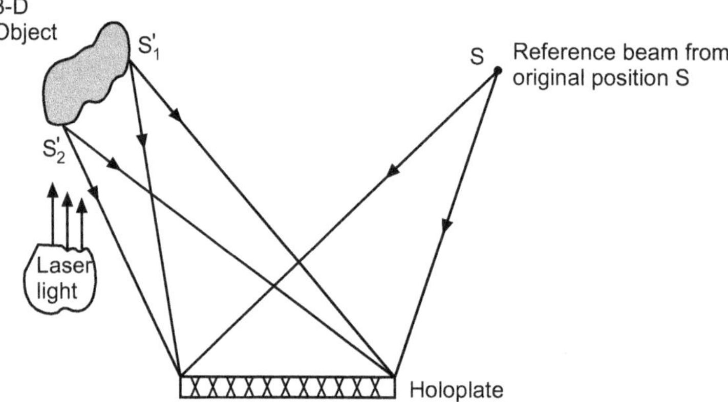

Fig. 7.7 : Generation of hologram

This missing key is provided later by shining a laser, identical to the one used to record the hologram, onto the developed film. When this beam illuminates the hologram, it is diffracted by the hologram's surface pattern. This produces a light field identical to the one originally produced by the scene and scattered onto the hologram.

(b) Super Market Scanner :

Laser bar code scanners are used in applications where the accurate identification of many articles in a short time is necessary, for example, in super markets, libraries and warehouses. These devices use a low powered laser to read the bar code attached to the article in question. A low power (He-Ne) laser beam is used to read the code. The super market scanner converts the pattern of bar code into a series of electrical signals. The laser beam is rapidly scanned in one or more planes usually by vibrating or rotating mirrors inside the bar code scanner. A sensor detects the time dependent pattern of reflected laser light as the beam is scanned across alternate light and dark bands on a bar code. This information is processed electronically to produce a digital interpretation of the bar code that a computer will understand. The information passes onto a central computer which identifies the product, looks up the price and prints the details on the bill. Bar codes are adopted now-a-days for a verity of purposes such as identifying materials in inventory of big concerns to maintain data on borrowed books in libraries, to quick process books and magazines by wholesale distributors.

(c) Compact Disc :

A compact disc or CD is an optical storage medium with digital data recorded on it. The digital data can be in the form of audio, video or computer information. When the CD is played, the information is read or detected by a tightly focused light source called a laser. CD's having a reflective aluminium layer that has very small pits put in the aluminium. The pits and the track of translated into binary by the computer and then are used for information. The storage of higher density of data is possible by using optical techniques. The storage medium is generally a thin film of metal whose optical properties change when it is illuminated with a powerful write laser. The less powerful read laser reads the change in optical property as the required information. Since laser beam can be focused on the spots smaller than one micrometer, it takes less than one square micro record one bit of information, i.e., 100 million per square cm.

Exercise

(A) Multiple Choice Questions :

1. ...... is an optical storage medium with digital data recorded.

   (a) compact disc　　　　(b) keyboard　　　　(c) hologram

2. A low power ...... beam is used to read the code.

   (a) He-Ne laser　　　　(b) Ruby laser　　　　(c) CO_2 laser

3. The record of ...... image of the object on a film is called a hologram.

   (a) 2D　　　　(b) 3D　　　　(c) 1D

4. ...... laser was used to weld detached retina.

   (a) Argon　　　　(b) CO_2　　　　(c) He-Ne

5. ...... lasers are range finding lasers.

   (a) Neodymium and CO_2　　(b) Argon and CO_2　　(c) Neodymium and He-Ne

6. ...... laser is used to separate it because it has very narrow linewidth.

   (a) Tuning dye　　　　(b) CO_2　　　　(c) He-Ne

7. For laser drilling, usually ...... laser is used.

   (a) Pulsed Ruby, CO_2 and He-Ne

   (b) pulsed carbon dioxide, Nd:YAG and Alexandrite

8. ...... laser used for cutting a wide range of materials.

   (a) He-Ne   (b) Ruby   (c) CO_2

9. ...... gases are often used in laser welding for protection against oxidation of metal surface.

   (a) He, Ar and N_2 (b) He, Ne and Ar   (c) Ne, Ar and CO_2

10. ...... laser is commonly used in welding of sheets.

   (a) He-Ne   (b) Ruby   (c) CO_2

Answers : (1) a, (2) a, (3) b, (4) a, (5) a, (6) a, (7) b, (8) c, (9) a, (10) c

(B) Short Answer Questions :

1. What are the applications of lasers ?
2. What is a hologram ?
3. State three types of holography.
4. Give various applications of lasers in medical field.
5. Explain various applications of lasers in mechanical industries.
6. Give various applications of laser beam in nuclear science field.
7. Give various applications of lasers in defence.
8. Give various applications of lasers in optical science field.

(C) Long Answer Questions :

1. What is a hologram ? Describe in short how a hologram is generated ?
2. State various uses of laser beam. Explain range finder application of laser in defense.
3. State various uses of laser beam. Explain eye surgery application of laser in medical.
4. Explain in brief nuclear isotope and nuclear fusion application of laser.
5. Write a short note on use of laser in welding of material.
6. Write a short note on use of laser in cutting of material.
7. Give various applications of laser in mechanical industries. Explain any one application.

❑❑❑

APRIL 2016

Time : 2 Hours]   **[Max. Marks : 40**

Instructions to the candidates :

    (1) All questions are compulsory.

    (2) Figures to the right indicate full marks.

    (3) Use of log table and calculator is allowed.

1. **Attempt all of the following (1 mark each) :**   **(10)**

    (a)   Who and when investigated First LASER.

    Ans. Please refer Chapter 1, Page 1.2, Article 1.2.

    (b)   State Boltzmann's equation an thermal equilibrium.

    Ans. Please refer Chapter 1, Page 1.6, Equation (1.7).

    (c)   State condition for population inversion.

    Ans. Please refer Chapter 2, Page 2.3, Article 2.3.

    (d)   Define metastable state in LASER.

    Ans. Please refer Chapter 2, Page 2.5, Article 2.4.

    (e)   Define round trip gain.

    Ans. Please refer Chapter 3, Page 3.3, Article 3.2.

    (f)   What is an optical feedback ?

    Ans. Please refer Chapter 3, Page 3.1, Article 3.1.

    (g)   Define gain bandwidth.

    Ans. Please refer Chapter 4, Page 4.5, Article 4.3.

    (h)   State two types of coherence.

    Ans. Please refer Chapter 5, Page 5.3, Article 5.4.

    (i)   State two types of liquid Lasers.

    Ans. Please refer Chapter 6, Page 6.8, Article 6.5.

    (j)   What is hologram ?

    Ans. Please refer Chapter 7, Page 7.7, Article 7.5.

2. **Attempt any two of the following (5 marks each) :**   **(10)**

    (a)   Explain Natural and Collision broadening.

    Ans. Please refer Chapter 4, Page 4.3 and 4.4, Article 4.2.1 and 4.2.1.

    (b)   Obtain Einstein coefficient relations in Laser action.

    Ans. Please refer Chapter 1, Page 1.7, Article 1.10.

    (c)   Explain the various characteristics of Laser beam in brief.

    Ans. Please refer Chapter 5, Pages 5.1 to 5.5, Articles 5.2, 5.3, 5.4, 5.5 (Take brief description).

3. **Attempt any two of the following (5 marks each) :**   **(10)**

    (a)   Calculate intensity of He-Ne laser beam having emissive power 1 mW and wavelength 6328 A° and intensity of ruby laser beam having emission power 5 mW and wavelength 6943 A°.

    Ans. Please refer Chapter 5, Page 5.7, Example 4.

(b) What will be the reflectivity of other cavity mirror if the reflectance of first mirror is 100% ? The length of the cavity is 15 cm and gain factor of LASER material is 0.0005 per cm.

Ans. Please refer Chapter 3, Page 3.7, Example 3.

(c) Calculate the wavelength of beam at which rate of spontaneous emission is equal to rate of stimulated emission for the temperature of 300°K.

Ans. Please refer Chapter 2, Page 2.8, Example 2.

4. (a) Attempt any one of the following : (8)

(i) State various uses of LASER beam. Explain range finder applications of LASER in defense.

Ans. Please refer Chapter 7 → Write the uses using all articles 7.1 to 7.5.

Range finder in detail : Please refer Chapter 7, Page 7.5, Article 7.3.

(ii) Describe construction and working of Ruby Laser. Also give its applications.

Ans. Please refer Chapter 6, Page 6.2, Article 6.2.1.

(b) Attempt any one of the following : (2)

(i) Define gain bandwidth.

Ans. Please refer Chapter 4, Page 4.5, Article 4.3.

(ii) Draw diagram for 4-level energy optical pumping.

Ans. Please refer Chapter 2, Page 2.7, Fig. 2.6.

NOVEMBER 2016

Time : 2 Hours]   **[Max. Marks : 40**

Instructions to the candidates :

(1) All questions are compulsory.

(2) Figures to the right indicate full marks.

(3) Use of log table and calculator is allowed.

1. Attempt all of the following (1 mark each) : (10)

(a) Who and when investigated First LASER.

Ans. Please refer Chapter 1, Page 1.2, Article 1.2.

(b) Define population density.

Ans. Please refer Chapter 1, Page 1.5, Article 1.5.

(c) Define metastable state in LASER.

Ans. Please refer Chapter 2, Page 2.5, Article 2.4.

(d) What are the various techniques of pumping in LASER.

Ans. Please refer Chapter 2, Page 2.5, Article 2.5.

(e) Define round trip gain.

Ans. Please refer Chapter 3, Page 3.3, Article 3.2.

  (f)   State the condition for critical population inversion.

Ans. Please refer Chapter 3, Page 3.4, Article 3.4.

  (g)   Define gain FWHM.

Ans. Please refer Chapter 4, Page 4.5, Article 4.3.

  (h)   State two types of coherence.

Ans. Please refer Chapter 5, Page 5.3, Article 5.4.

  (i)   State any two types of gas lasers.

Ans. Please refer Chapter 6, Page 6.4, Article 6.3.

  (j)   State various applications of LASER beam in mechanical industries.

Ans. Please refer Chapter 7, Page 7.1, Article 7.1.

2.　Attempt any two of the following (5 marks each) :　　　　　　　　**(10)**

  (a)   Explain the various characteristics of LASER beam in brief.

Ans. Please refer Chapter 5, Pages 5.1 to 5.5, Article 5.2, 5.3, 5.4, 5.5 (Take brief description).

  (b)   What is a Line shape broadening ? Explain homogeneous and inhomogeneous broadening.

Ans. Please refer Chapter 4, Page 4.2, Article 4.2.

  (c)   Obtain Einstein's coefficient relations in LASER action.

Ans. Please refer Chapter 1, Page 1.7, Article 1.10.

3.　Attempt any two of the following (5 marks each) :　　　　　　　　**(10)**

  (a)   What will be the reflectivity of other cavity mirror if the reflectance of first mirror is 100% ? The length of the cavity is 15 cm and gain factor of LASER material is 0.0005 per cm.

Ans. Please refer Chapter 3, Page 3.7, Example 3.

  (b)   The half – width of gain profile of He-Ne laser material device is 0.002 nm having length of cavity is 10 cm. Calculate emitted wavelength of laser in order to single mode of oscillation having refractive index of material is 1.

Ans. Please refer Chapter 4, Page 4.6, Example 1.

  (c)   Light from a He-Ne laser, which is the traditional monochromatic source, has coherence length of about 100 m and wavelength 6328 A°. Calculate bandwidth of He-Ne laser.

Ans. Please refer Chapter 5, Page 5.6, Example 3.

4.　(a)　Attempt any one of the following :　　　　　　　　**(8)**

  (i)   State construction and working of Ruby LASER. Also give its applications.

Ans. Please refer Chapter 6, Page 6.2, Article 6.2.1.

  (ii)   What is a hologram ? Describe in short how hologram is generated.

Ans. Please refer Chapter 7, Page 7.7, Article 7.5.

(b) Attempt any one of the following : (2)

(i) Draw the block diagram for optical resonator and label it.

Ans. Please refer Chapter 3, Page 3.4, Article 3.5.

(ii) Draw diagram for 4-level optical pumping and label it.

Ans. Please refer Chapter 2, Page 2.7, Fig. 2.6.

APRIL 2017

Time : 2 Hours] **[Max. Marks : 40**

Instructions to the candidates :

(1) All questions are compulsory.

(2) Figures to the right indicate full marks.

(3) Use of log table and calculator is allowed.

1. Attempt all of the following (1 mark each) : (10)

(a) Define population density.

Ans. Please refer Chapter 1, Page 1.5, Article 1.5.

(b) What is laser light ?

Ans. Light which consists of the radiations of only one wavelength is called laser light. It is highly directional and travels only in one direction. It is highly coherent and can be focused to a very small spot.

(c) Define active medium in lasers.

Ans. Please refer Chapter 2, Page 2.5, Article 2.4.

(d) What is pumping ?

Ans. Please refer Chapter 2, Page 2.5, Article 2.5.

(e) Define round trip gain.

Ans. Please refer Chapter 3, Page 3.3, Article 3.2.

(f) What is hologram ?

Ans. Please refer Chapter 7, Page 7.7, Article 7.5.

(g) What is optical feedback ?

Ans. Please refer Chapter 3, Page 3.1, Article 3.1.

(h) Write two applications of solid state lasers.

Ans. Please refer Chapter 6, Page 6.2, Article 6.2.

(i) Which active medium is used in tunable dye lasers ?

Ans. In tunable dye lasers, an organic dye (liquid solution) is used as an active medium. It consists of a host material such as alcohol in which the dye molecules (such as rhodamines or coumarins) are dissolved at a concentration of one part in ten thousand.

(j) What is range finder ?

Ans. Please refer Chapter 7, Page 7.5, Article 7.3.

2. Attempt any two of the following (5 marks each) :   **(10)**

   (a) Explain characteristics of lasers.

Ans. Please refer Chapter 5, Page 5.1 to 5.5, Articles 5.2, 5.3, 5.4, 5.5 (Take brief description).

   (b) Distinguish between ordinary light and laser light.

Ans. Please refer Chapter 1, Page 1.1, Article 1.1.

   (c) What is line broadening ? Explain Doppler broadening.

Ans. Please refer Chapter 4, Page 4.2, 4.5, Article 4.2, 4.2.3.

3. Attempt any two of the following (5 marks each) :   **(10)**

   (a) Fluorescent tube light emits visible light of wavelengths in the range of 4000 A° to 7000 A° with average wavelength of 5500 A°. Calculate coherence length and time.

Ans. Please refer Chapter 5, Page 5.6, Example 1.

   (b) Find the ratio of population of two states in a He-Ne laser, that produces light of wavelength 6328 A° at 27°C.

Ans. Please refer Chapter 1, Page 1.10, Example 1.

   (c) Calculate length of cavity and optical path for He-Ne laser tube which achieves the condition for amplification having wavelength 6328 A° and refractive index of active medium between two neighbouring reflectors is 1.56.

Ans. Please refer Chapter 3, Page 3.6, Example 2.

4. (a) Attempt any one of the following :   **(8)**

   (i) Explain various applications of lasers in mechanical industries.

Ans. Please refer Chapter 7, Page 7.1, Article 7.1.

   (ii) Describe construction and working of He-Ne laser.

Ans. Please refer Chapter 6, Page 6.4, Article 6.3.1.

(b) Attempt any one of the following :   **(2)**

   (i) Define energy levels.

Ans. Please refer Chapter 1, Page 1.4, Article 1.4.

   (ii) Draw the block diagram for optical resonator.

Ans. Please refer Chapter 3, Page 3.4, Article 3.5.

OCTOBER 2017

Time : 2 Hours]   **[Max. Marks : 40**

Instructions to the candidates :

   (1) All questions are compulsory.

   (2) Figures to the right indicate full marks.

   (3) Use of log table and calculator is allowed.

1. Attempt all of the following (1 mark each) :   **(10)**

   (a) Define population density.

Ans. Please refer Chapter 1, Page 1.5, Article 1.5.

(b) Who and when investigated First MASER ?

Ans. In 1951, Charles H. Townes and his co-workers independently investigated and suggested the principle of First MASTER (Microwave Amplification of Stimulated Emission of Radiation) based on stimulated emission, at Columbia University and Lebedev Institute of Physics, Moscow.

(c) Define metastable state in LASER.

Ans. Please refer Chapter 2, Page 2.5, Article 2.4.

(d) What are the various techniques of pumping in LASER ?

Ans. Please refer Chapter 2, Page 2.5, Article 2.5.

(e) What is an optical feedback ?

Ans. Please refer Chapter 3, Page 3.1, Article 3.1.

(f) State the condition for critical population inversion.

Ans. Please refer Chapter 3, Page 3.4, Article 3.4.

(g) What is beam radiance ?

Ans. The light emitted by a laser is confined to a rather narrow cone. But, when the beam propagates outward, it slowly diverges or fans out. For an electromagnetic beam, beam divergence is the angular measure of the increase in the radius or diameter with distance from the optical aperture as the beam emerges.

OR

The beam divergence is a measure for how fast a laser beam expands far from its focus.

(h) Define FWHM.

Ans. Please refer Chapter 4, Page 4.5, Article 4.3.

(i) State any two types of gas lasers.

Ans. **Gas lasers :** He-Ne laser, CO_2 laser.

(j) What is hologram ?

Ans. Please refer Chapter 7, Page 7.7, Article 7.5.

2. **Attempt any two of the following (5 marks each) :**   **(10)**

(a) What do you understand by coherence ? Explain temporal and spatial coherence ?

Ans. Please refer Chapter 5, Page 5.3, Article 5.4.

(b) What is lineshape broadening ? Explain collision broadening.

Ans. Please refer Chapter 4, Page 4.2, 4.4, Articles 4.2, 4.4.

(c) Explain basic three processes of LASER with neat diagram.

Ans. Three basic processes of Laser are Absorption, spontaneous emission and Stimulated emission.

For explanation please refer Chapter 1, Page 1.7, Article 1.9.

3. **Attempt any two of the following (5 marks each) :**   **(10)**

(a) Calculate the wavelength of beam at which rate of spontaneous emission is equal to rate stimulated emission for the temperature of 300°K.

Ans. Please refer Chapter 2, Page 2.8, Example 2.

  (b)   What will be the reflectivity of first cavity mirror if the reflectance of second mirror is 97%. The length of the cavity is 15 cm and gain factor of laser material is 0.0005 per cm.

Ans. Please refer Chapter 3, Page 3.7, Example 4.

  (c)   Calculate intensity of He-Ne laser beam having emissive power 1 mW and wavelength 6328 A° and intensity of ruby laser beam having emissive power 5 mW and wavelength 6943 A°.

Ans. Please refer Chapter 5, Page 5.7, Example 4.

4. (a)   Attempt any one of the following :   **(8)**

  (i)   State various uses of LASER beam. Explain range finder applications of Laser in defense.

Ans. Please refer Chapter 7 → Write the uses using all articles 7.1 to 7.5.

     Range finder in detail : Please refer Chapter 7, Page 7.5, Article 7.3.

  (ii)   Describe construction and working of He-Ne Laser. Also give its applications.

Ans. Please refer Chapter 6, Page 6.4, aRticle 6.3.1.

(b)   Attempt any one of the following :   **(2)**

  (i)   Draw diagram for 4-level energy optical pumping.

Ans. Please refer Chapter 2, Page 2.7, Fig. 2.6.

  (ii)   Define gain bandwidth.

Ans. Please refer Chapter 4, Page 4.5, Article 4.3.

❑❑❑

APRIL 2018

Time : 2 Hours]   **[Max. Marks : 40**

Instructions to the candidates :

  (1) All questions are compulsory.

  (2) Figures to the right indicate full marks.

  (3) Use of log table and calculator is allowed.

1. Attempt all of the following (1 mark each) :   **(10)**

  (a)   Define metastable state.

Ans. Please refer Chapter 2, Page 2.5, Article 2.4.

  (b)   What is difference between laser light and normal light ?

Ans. Please refer Chapter 1, Page 1.1, Article 1.1.

  (c)   State any two types of lasers.

Ans. Please refer Q. 1 (ii) of October 2017.

  (d)   What is population inversion ?

Ans. Please refer Chapter 2, Page 2.3, Article 2.2.

  (e)   Give two types of liquid lasers.

Ans. Types of liquid lasers : Tunable dye laser, Excimer laser.

  (f)   What is holography ?

Ans. Please refer Chapter 7, Page 7.7, Article 7.5.

(g)   Define spontaneous emission.

Ans. Please refer Chapter 1, Page 1.7, Article 1.9.

(h)   Define active medium in laser.

Ans. Please refer Chapter 2, Page 2.5, Article 2.4.

(i)   What is pumping ?

Ans. Please refer Chapter 2, Page 2.5, Article 2.5.

(j)   Give applications of lasers in defence.

Ans. Please refer Chapter 7, Page 7.5, Article 7.3.

2. Attempt any two of the following (5 marks each) :   **(10)**

(a)   Write a short note on optical feedback.

Ans. Please refer Chapter 3, Page 3.1, Article 3.1.

(b)   Explain the characteristics/properties of a laser beam.

Ans. Please refer Chapter 5, Page 5.1 to 5.5, Article 5.2, 5.3, 5.4, 5.5 (Take brief description).

(c)   Give two techniques of pumping and explain three-level pumping scheme.

Ans. Two techniques of pumping : (i) Three-level optical pumping scheme and (ii) Four-level optical pumping scheme.

Three-level pumping scheme : Please refer Chapter 2, Page 2.6, Article 2.5.

3. Attempt any two of the following (5 marks each) :   **(10)**

(a)   Light from a sodium lamp, which is the traditional monochromatic source, has coherence length of about 0.3 mm and bandwidth of about 6 A°. Calculate wavelength of sodium light.

Ans. Please refer Chapter 5, Page 5.6, Example 1.

(b)   What will be the reflectivity of other cavity mirror if the reflectance of first mirror is 100% ? The length of the cavity is 15 cm and gain factor of laser material is 0.0005 per cm.

Ans. Please refer Chapter 3, Page 3.7, Example 3.

(c)   Find the ratio of population of the two states in a He-Ne laser that produces light of wavelength 6328 A° at 27°C.

Ans. Please refer Chapter 1, Page 1.10, Example 1.

4. (a) Attempt any one of the following :   **(8)**

(i)   Explain the applications of lasers in mechanical industry.

Ans. Please refer Chapter 7, Page 7.1, Article 7.1.

(ii)   Explain with neat diagram construction and working of CO_2 laser. Also give its applications.

Ans. Please refer Chapter 6, Page 6.5, Article 6.3.2.

(b) Attempt any one of the following :   **(2)**

(i)   Draw diagram of He-Ne laser.

Ans. Please refer Chapter 6, Page 6.4, Fig. 6.4.

(ii)   What is stimulated emission ?

Ans. Please refer Chapter 1, Page 1.7, Article 1.9.

❑❑❑

www.ingramcontent.com/pod-product-compliance
Lightning Source LLC
LaVergne TN
LVHW080436200726
843507LV00004B/843